PHYSICS RESEARCH AND TECHNOLOGY

THEORETICAL BASIS AND MATHEMATICAL MODELING OF THE THERMAL NEUTRON CONVERTERS BASED ON GADOLINIUM FOILS

PHYSICS RESEARCH AND TECHNOLOGY

Additional books and e-books in this series can be found on Nova's website under the Series tab.

PHYSICS RESEARCH AND TECHNOLOGY

THEORETICAL BASIS AND MATHEMATICAL MODELING OF THE THERMAL NEUTRON CONVERTERS BASED ON GADOLINIUM FOILS

DJAMSHED A. ABDUSHUKUROV

nova
science publishers
New York

Library of Congress Cataloging-in-Publication Data

ISBN: 978-1-53614-225-9

Published by Nova Science Publishers, Inc. † New York

CONTENTS

PREFACE

Converters of neutron radiation play a determining role in the development of detectors of these radiations. They define the basic characteristics of detectors: the efficiency of registration, energy, time and spatial resolution.

Among the solid-state converters the gadolinium and it's 157 isotopes are especially allocated, possessing abnormal high cross section of interaction with thermal neutrons.

In the book, theoretical bases of the registration of neutron radiation by converters from gadolinium are considered. The efficiency of converters is the product of three variables. These are the following: (1) Probabilities of capture of thermal neutrons by nucleus; (2) Probabilities of creation the secondary charged particles, in our case the internal conversion and Auger electrons; (3) Probabilities of escape of created electrons from the material of the converter.

Model calculations of registration efficiency of thermal neutrons by the foil converters made from natural gadolinium and its 157 isotope described. Processes of neutron absorption in the material of a converter and the probability of secondary electron escapes examined. Calculation made for converters with the various thicknesses, and other parameters of converters. It was chosen the most optimal converter thicknesses.

The contribution low-energy Auger electrons radiated from L- subshell with the energy 4.84 keV and M-subsell with the energy 0.97 keV on efficiency of converters are lead. These electrons have rather small free path length in gadolinium; these are 0.3 microns (4.84 keV) and 0.04 microns (0.97 keV). But their contribution to become essential at use of converters from 157 gadolinium isotope as the length of free path of neutrons in them does not exceed 2-3 microns and this length to become comparable with length of electrons path.

Estimation of contribution of X-rays and low-energy gamma-quantas absorbed directly in the converter and resulting of occurrence of secondary electrons was made. In case of the account of the contribution of electrons formed by X-ray quantas the efficiency is increased a little, but their contribution is no more, than by 1%.

Calculations of complex converters representing as two thin gadolinium foil located on the both sides of supporting kapton foils and complex converters representing a set thin foils drilled with the fine step located one over other in a one gas volume are presented.

Examples of development of detectors of thermal neutrons based on gadolinium converters described.

D. A. Abdushukurov

Chapter 1

INTRODUCTION

Mechanisms for detecting neutrons in matter are based on indirect methods. Neutrons are neutral, they do not interact directly with the electrons in matter. The neutrons can caused in a nuclear reaction. The process of neutron detection begins when neutrons, interacting with various nuclei, initiate the release of one or more charged particles. The products from these reactions, such as protons, alpha particles, gamma rays, electrons and fission fragments, can initiate the detection process. For detection of neutrally charged neutrons, it is necessary to use converters for neutron radiation. These converters will converse radiations of neutrons to the charged radiation, which can be in further detected with manifold detectors.

For an optimal choice of the converter foil material, the following properties should considered:

- A large neutron-capture cross section, needed in order to achieve high detection efficiency with minimum foil thickness.
- The range of the neutron-induced charged primary particles should be large compared with the convertor foil thickness.
- The converter material should possess small cross-section of activation. Daughter isotopes occurring as result of activation should not create the significant background.
- The γ-ray sensitivity of the material should be low.

At the present time variety detectors with gaseous and solid-state neutron converters are used. Only several kinds of nuclear reactions can used in these converters. The most widely known are ^{3}He and $^{10}BF_3$ based converters [1]. Gaseous converters used in the wide class of gas detectors. Solid-state converters applied in gaseous, semiconductor and in scintillation detectors.

In the development of gaseous detectors of thermal neutrons the highest efficiency of registration up to 80% have been reached to receive by the uses of gas mixtures on the basis of Helium - 3 (^{3}He) at pressure up to 15 atm. The use of the high pressure is necessary to increase the efficiency of registration and improvement of the spatial resolution. The increase of pressure dictate the additional requirements to the mechanical designs and first of all to the thickness of the entrance windows. At the same time, the high price of the gas and fluidity of ^{3}He results in sharp rise of price of detectors and their operation.

Converters of neutron radiation play a determining role at designing detectors of neutron radiation. They define the key parameters of detectors, such as the efficiency, the spatial resolution and so on. Among the solid-state converters of thermal neutrons the highest efficiency of registration has received by using the gadolinium based converters, and especially, its 157 isotope. As a result of capture of thermal neutrons by nucleus of gadolinium the internal conversion electrons and Auger electrons are radiated. Another solid-state converter rarely used in detectors of neutrons, which is first of all connected with the low efficiency of registration. Therefore, the efficiency of registration of detectors based on ^{10}B and ^{6}Li usually do not exceed 3-4% and 1%, respectively [2, 3].

The idea of using thin foils to convert neutrons radiation into charged particles and count the conversion products in a solid-state detector was originally proposed by Feigl and Rauch [4, 5]. They used natural Gd and pure ^{157}Gd convertor foils and measured of the escape probabilities and the energy distributions of the escaping electrons with Si surface barrier detectors.

In the paper A. P. Jeavons et al. [6], the efficiency of registration of thermal neutrons by the converters constructed on base of gadolinium foils has been simulated. The following conditions were considered at the modeling: the neutron beams (2 fixed energies) perpendicularly fell on to the converting foils of various thicknesses; the output of the secondary electrons has been calculated both in the lobby and the back hemispheres. As result the detector on the basis of the multiwire proportional chamber with the gadolinium converter has been developed. Detector did not have enough good spatial resolution about 10 mm. In order to improve the spatial resolution, the same authors have offered to use the converter - collimator, which was the sandwich created from lead foils with the fiberglass layers and drilled with the fine step (Jeavons converter). This converter was nestled to the gadolinium foil and thus limited electron runs in the gas, simultaneously serving as the emitter of the secondary electrons. It has helped to improve the spatial resolution to 2 mm, but the spatial resolution turned to modulation with step of holes, besides the efficiency of registration has sharply fallen.

Further, in G. Charpak group [7], during the development of multistep avalanche chambers with gadolinium converters it was shown sharply improve of the spatial resolution of detectors (up to 1 mm) without use of Jeavons converter [8].

In the paper D. A. Abdushukurov et al. [9-11], mathematical modeling of the efficiency of registration of thermal neutrons by the gadolinium foils has been conducted. In contrast to earlier works the incidence angle of neutron beam was not a constant value (90 degrees) and varied from 1 up to 90 degrees. As the result the conclusion about the increase of efficiency of registration of neutrons at the small angles (less 10 degrees) between the converter and beam of incidenced neutrons has been done. It is connected with the increase of neutron path length in the body of the converter at conserving of effective thickness of material for the output of secondary electrons. At the disposition of two detectors on the different sides of the converter, the efficiency of registration of thermal neutrons can increased up to 60%. In the carried out calculations, we took into account only electrons with the energies higher than 29 keV. Thus only electrons of internal conversion and Auger electrons, radiated from K-subshell, with the energy 34.9 keV [12]. The minimal free path length of the electrons with energy 29 keV in gadolinium makes 4.7 microns. The similar choice of energies of electrons was made not only by us, but also and by other researchers.

Later we done calculations of influence of low-energy secondary electrons which are radiated from gadolinium foils. These are Auger electrons radiated from a L-subshell with the energy 4.84 keV and electrons from M-subshell with energy 0.97 keV. These electrons have rather small free path length in gadolinium 0.3 microns (4.84 keV) and 0.04 microns (0.97 keV), thus accordingly make average free path length of these electrons 0,1 and 0,015 microns. During calculations it was found out, that contribution of low energy electrons becomes essential at the use of converters made from 157 isotope of gadolinium, as the length of free path of neutrons in gadolinium does not exceed 2-3 microns and this length becomes comparable with the free path length of electrons [13]. Results of modeling calculations within the limits of errors have coincided with the experimental data published in [14]. The good consent of the calculated data with the experimental ones testifies to correctness of the chosen models and theoretical preconditions.

In the literature and the tabulated data bases there are no data for the free path length of electrons with the energy less than 10 keV. All available data for electrons in various materials received from the Bethe-Bloch theory. The lower border of applicability of this theory is the energy of 10 keV. The theory constructed on the assumption, that the charged particles have continuous losses of energy by two mechanisms at their motion in materials. These are losses of energy on irradiations and direct collisions. For electron with energies less than 10 keV the new mechanism added, it is the probability of capture of free electrons by the atom subshells having vacancies. Also practically, there are no data on the quantity of low-energy Auger electrons. Available data differs with an error more than 100%. During realization of calculations, it is necessary to compare received results with the experimental data. It will allow estimating the correctness of theoretical preconditions and modeling representations.

The thin layer converters will allow creating new types of detectors of thermal neutrons having the improved characteristics. Therefore, it will be possible to develop detectors with pico-second time resolution for time-resolved experiments, also an improvement of the spatial resolution can expected due to the reduction of thickness and parallax of detectors. It is possible to specify an opportunity of miniaturization of detectors materials that claimed for many applications.

Several research groups have been engaged in mathematical modeling of the characteristics of gadolinium converters. The results obtained are well described in the work of the Pfelffer et al. [15]. Below is a citation from her article.

> A number of 59.9 conversion electrons per 100 captured neutrons in natural Gd can be found in the work of Harms and McCormack [16], whereas Hargrove et al. based on the gamma level schemes in the work of Bäcklin et al. [17] and Greenwood et al. [18] give the theoretical number of conversion electrons per captured neutron as 0.66 [19]. To calculate the neutron detection efficiency of the detector, not the number of conversion electrons per captured neutron is relevant, but the probability that at least one conversion electron is produced and reaches the drift. The probability for the production of at least one conversion electron in the converter amounts to 51.1%. This number is lower than the number of conversion electrons per 100 neutrons, because in some cases more than one conversion electron is produced during the neutron capture. In backwards mode, the probability of at least one conversion electron reaching the drift space is 14.2%. Assuming that each conversion electron that reaches the drift can be detected, the detection efficiency of the detector is 14.2% in backwards configuration. Taking into consideration the scattering, the detection efficiency is 0.813 times 0.142 or 11.5%.
>
> Bruckner et al. calculated the detection efficiency for a Si detector with a natural Gd converter of 25 μm thickness in backwards configuration [14]. For neutrons with 1.8Å wavelength, the calculated efficiency was around 11.6%. According to Abdushukurov, the calculated detection efficiency for a 100 μm thick natural Gd converter in backwards mode with neutrons of 1.8 Å wavelength amounts to 15.2% [20]. With 1.8 Å neutrons and without scattering of the neutrons, Geant4 arrives in backwards configuration at a detection efficiency of 13.9% for a converter thickness of 25 μm and 14.1% for a converter thickness of 100 μm. The simulated efficiencies are thus slightly lower than the efficiency calculated by Abdushukurov and slightly higher than the efficiency calculated by Bruckner et al., but agree reasonably well with the literature. In general, there seems to be the tendency that measured efficiencies are higher than calculated efficiencies. The measured efficiency by Bruckner

et al. for various wavelengths was considerably higher than their calculation, e.g. for 1.7 Å neutrons 19% were measured with a 25 μm thick converter. The authors explain the discrepancy with the partial detection of the X-rays and gamma rays that are produced during the neutron capture. Hargrove et al. derive from their measurements with thermal neutrons an impressive total efficiency (backwards and forwards mode) of 63% for a 4 μm thick foil of natural Gd. They try to explain the discrepancy with contributions from higher gamma levels that are usually not included in the theoretical calculations of the conversion electron spectrum.

Computer modeling allows to prospect, without special material inputs of the most suitable configuration of converters, to carry out search of optimum geometrical ratio of convertors. It in turn allows refusing of realization of superfluous development and researches.

In present time the position - sensitive detection of thermal neutron radiation widely applied in various detectors with solid-state converters such as a normal-pressure multistep avalanche chambers (MSAC), low-pressure multistep avalanche chambers (LMSAC), microchannel plates, thin layer scintillators. New types of detectors are offered and developed such as hybrid low-pressure micro-strip gas chamber (MSGC), position sensitive silicon detectors (Gd Si PD), resistive plate chambers (RPCs), imaging plate neutron detectors (IP-NDs), GEM structure, liquid scintillator into a capillary plate and so on.

Chapter 2

MATHEMATICAL MODELING OF THIN LAYER NEUTRON CONVERTERS

2.1. THEORETICAL BASES

In the current chapter influence of various parameters of converters on their efficiency are viewed. Efficiency is our understanding is the ration of the electrons which have departed from converters, to total number of falling neutrons on the converter. The efficiency of converters is the product of three variables. These are the following:

- Probabilities of thermal neutrons capture by nucleus of the converter, which depends on its thickness, and the cross-section of interaction. The cross-section of interaction in turn depends on isotope composition of the converter and energy or wavelength of neutron.
- Probabilities of creation of the secondary charged particles, in our case of internal conversion and Auger electrons.
- Probabilities of escape of the created electrons from the material of the converter, which depends on, free path length of electrons in a converter material and geometry of emission.

Energies of the secondary electrons have a discrete values and are formed with the certain probability. At calculation of probability of escape, it is necessary to take into account emission of each electron with its characteristic energy and its weight factor (probability of formation). In that specific case efficiency can be described by the following expression

$$E_i = P_n(\sigma L_j) * D_e\,(E_i N_i) * H_e(Re_i \Theta_i) \qquad (1)$$

where E_i is the efficiency of registration of the absorbed with the probability $P_n(\sigma L_j)$ neutrons, which caused formation of electrons $D_e(E_i N_i)$ with the discrete energies and probability of formation, probability of escape of electrons from a material of converter $H_e(Re_i \Theta_i)$ in view of the maximal free path length of electrons in a material of the converter and angle of their emission. Generally, probabilities are necessary for summarizing, so calculations are made for 446 discrete energies of electrons with the energies in the range from 0.9 up to 1000 keV.

$$E = \Sigma\Sigma\Sigma(P_n(\sigma L_j) * D_e\,(E_i N_i) * H_e(Re_i \Theta_i)) \qquad (2)$$

2.1.1. Probability of Neutrons Absorption

The attenuation of the narrow collimated neutron beam in thin layer materials is governed by the exponential law [21]

$$F_x = F_0 \exp\,(-N_A\, \sigma\, X) \qquad (3)$$

where F_x and F_0 are the neutron flux density after and before its passage through the layer of the material with the thickness X, correspondingly, N_A is the number of nucleus in the volume of 1 cm^3, σ- is full microscopic cross-section of neutron interaction with the nuclei of material.

One can simplify the formulae for the neutron flux density on the distance R (cm or g/cm^2) from the dot isotropic neutron source, emitting I_0, neglecting the exponential correction on attenuation when $\lambda \leq\leq R$ [22]

$$F_R = I_0/4\,\pi\, R^2 \qquad (4)$$

In our calculations we use four fixed neutron energies. These are neutrons with wavelengths of 1, 1.8, 3 and 4 Å. In Table 1 their wavelengths, corresponding energies (eV) and velocities (m/s) are shown.

Natural gadolinium is a mixture of isotopes that could participate in the (n,γ) nuclear reaction. The basic characteristics of the most widespread isotopes gadolinium including cross section of interaction with neutrons, daughter isotopes, and a half-life period of unstable daughter isotopes, are presented in Table 2 [23, 24].

As one can see in this table, the most interesting for our calculations are natural Gd and its 155 and 157 isotopes, which have abnormally high cross sections of interaction with neutrons. Other isotopes give an insignificant contribution to the interaction with neutrons.

Table 1.

Wavelength Å	Energy of neutrons (eV)	Velocity of neutrons (m/c)	Capture Cross Section (barns) for ^{Nat}Gd	Capture Cross Section (barns) for ^{157}Gd
1	0,081894	3955	13 563.56	75 323.47
1,8	0,025276	2197,2	48 149.41	253 778.40
3	0,0090993	1318,3	70 597.77	367 842.60
4	0,0051184	988,76	89 066.84	464 373.40

Table 2.

Isotope	Abundance (%)	Cross section (b)	Daughter isotope	$T_{1/2}$
^{nat}Gd	100	48890	-	-
^{152}Gd	0.2	1100	^{153}Gd	241.6 d
^{154}Gd	2.2	90	^{155}Gd	Stable
^{155}Gd	14.7	61000	^{156}Gd	Stable
^{156}Gd	20.6	2.0	^{157}Gd	Stable
^{157}Gd	15.68	255000	^{158}Gd	Stable
^{158}Gd	24.9	2.4	^{159}Gd	18.6 h
^{160}Gd	21.9	0.8	^{161}Gd	3.66 min

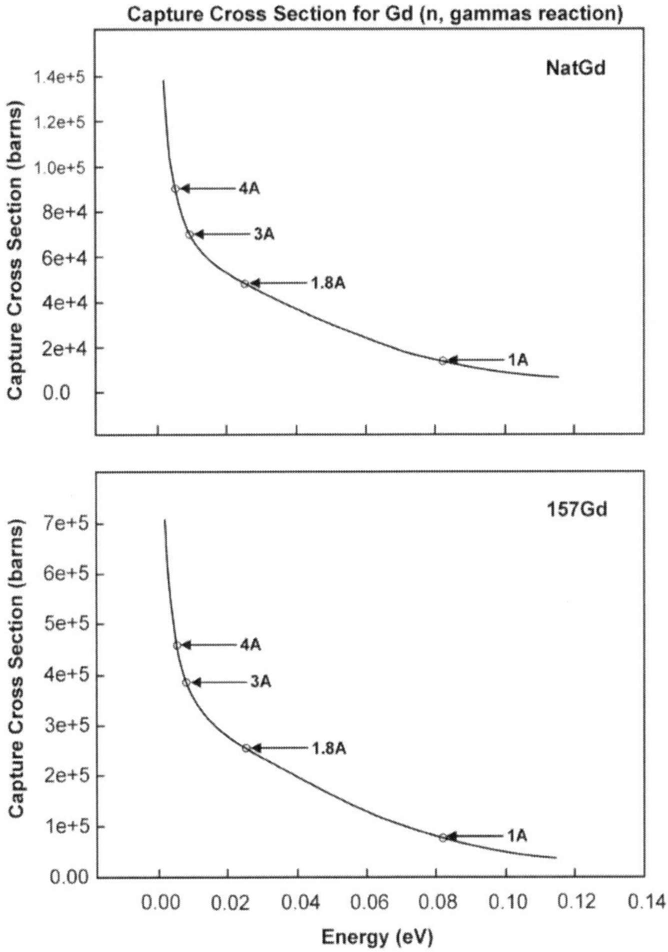

Figure 1. Cross-section of thermal neutrons capture, for the reaction (n,γ), depending on energy of neutrons for natural gadolinium and its 157 isotope [25].

In Figure 1 the dependence of cross-section (barn) on the energy of incident neutrons (eV), for natural gadolinium and the same dependence for its 157 isotope are shown. Arrows indicate energy of neutrons for which we will carry out our calculations. As one can see, with the reduction of energy of neutrons the cross-section of interaction strongly increases. It especially increases in the region of cold and ultra-cold neutrons.

Table 3.

Isotope	Daughter isotope	Eγ [kev]	Cross section [b] (error)	Iγ [1/100 n] (error)
157-Gd	158-Gd	79.510	4010(100)	77.3(19)
157-Gd	158-Gd	135.26	38(4)	0.73(8)
157-Gd	158-Gd	181.931	7200(300)	139(6)
157-Gd	158-Gd	212.97	10.8(7)	0.21(13)
157-Gd	158-Gd	218.225	55(4)	1.06(8)
157-Gd	158-Gd	230.23	20.0(11)	0.385(21)
157-Gd	158-Gd	255.654	350(19)	6.7(4)
157-Gd	158-Gd	277.544	493(12)	9.50(23)
157-Gd	158-Gd	365	59(5)	1.14(10)
157-Gd	158-Gd	780.14	1010(22)	19.5(4)
157-Gd	158-Gd	944.09	3090(70)	59.5(13)
157-Gd	158-Gd	960	2050(130)	39.5(25)
157-Gd	158-Gd	975	1440(21)	27.8(4)

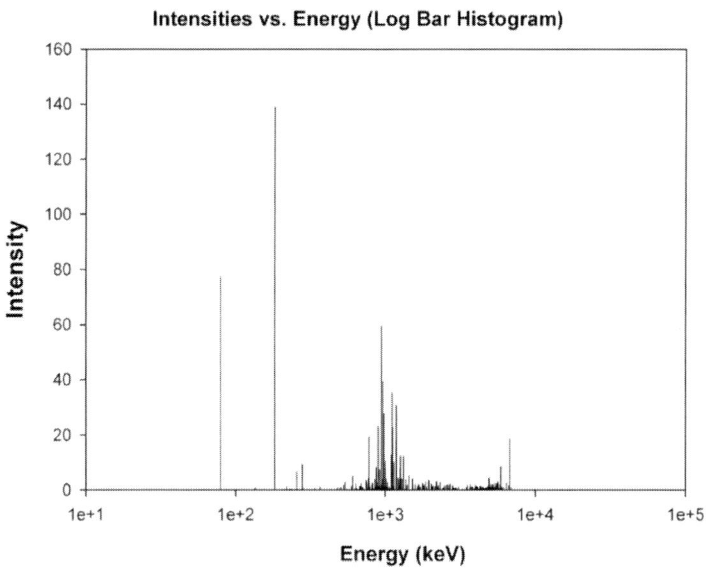

Figure 2. Histogram of dependence of intensity of gamma-quanta on energy for the reaction of ^{157}Gd (n, γ) ^{158}Gd (by 100 neutrons).

2.1.2. Probability of Gamma Quanta's Formation

In the reaction of ^{157}Gd neutron capture, 7937.33 keV energy is emitted. In total 390 lines with energy ranges from 79.5 up to 7857.670 keV with line intensity of 2×10^{-8} up to 139 gamma-quanta on 100 captured neutrons are emitted. In Table 3 the most intensive, low-energy gamma-lines having high coefficient of internal conversion are presented [26]. In Figure 2 the histogram showing dependence of gamma quantum intensity on the energy is presented.

2.1.3. Probability of Internal Conversion Electrons Formation

In gamma-quantum spectrum the low-energy quantas are presented. During their emission the electrons from atomic subshells (so called internal conversion electrons) can radiated with a high probability. The nuclear removes from its excitation by radiating a gamma-quantum, but also a close located electron from electron shells can be irradiated. Usually K-electron (electron from K-subshell) is emitted, but also electrons from the higher subshells (like L, M, N and so on) can be emitted too. Vacancy of electrons (an electronic hole), formed as a result of this process, is filled by another electron from a higher level. This process is accompanied by radiation of X-ray quantum, or radiation of Auger-electron.

The probability of formation of internal conversion electrons can consider as follows. An electromagnetic decay of the atomic nucleus can proceed by competing modes: electromagnetic radiation (γ), production of electron-positron pairs (e^+e^-) or emission of orbital electrons (e^-) that is, internal conversion. The conversion coefficient α is the ratio of the electron emission rate (T_e) to the gamma emission rate (T_γ),

$$\alpha = T_e/T_\gamma \quad (5)$$

The values of α are depend on four parameters: (1) the charge of the decaying nucleus; (2) the energy of the nuclear transition; (3) the atomic subshell of which the orbital electron is ejected; (4) the multi-polarity and parity of the nuclear transition. The knowledge of the coefficients is one of the most important tools for the determination of parity and multipolarity of electromagnetic nuclear transitions and the construction of nuclear decay schemes, but other applications exist as well.

The emission of nuclear gamma rays is accompanied by the emission of orbital electrons. Their branching ratio is the conversion coefficient α. The discovery of this process and its naming as a "conversion of the γ-radiation" is due to Hahn and Meitner [27]. The first correct theoretical description is due to Hulme in 1932. A review of the theory was given after by Pauli, Alder, and Steffen [28]. In the lowest, non-trivial order of perturbation theory the conversion coefficient for a transition of pure electric multipole order L is given by

$$\alpha_\sigma(EL) = \pi\alpha\omega \sum_k \frac{(2j+1)(2j+1)}{L(L+1)} \begin{pmatrix} j_0 & j & L \\ 1/2 & -1/2 & 0 \end{pmatrix}^2 \times \left|R_{kk}(EL) + T_{kk_0}(EL)\right|^2 \tag{6}$$

The index σ refers to the atomic orbital (subshell) out of which the electron is ejected. The quantities in big parentheses are 3-*j* symbols. The total conversion coefficient is given by

$$\alpha_t = \sum_i \alpha_{\sigma_i}, \tag{7}$$

and the total electromagnetic decay rate by

$$T = T_\gamma + T_e = T_\gamma(1 + \alpha_t) \tag{8}$$

for transition energies $\omega < 2m_e c^2$.

The relativistic angular-momentum quantum number is related to the total angular momentum *j* by $j = |k| - 1/2$. An index zero refers to the initial bound state. A subshell σ is represented in addition by n_0, the principal quantum number, i.e., σ=(n_0, k_0). The so-called dynamic radial matrix elements T_{kk_0} contain nuclear transition currents and charges,

$$T_{kk_0} = 0. \tag{9}$$

For later references introduced a few other quantities. The bound energy of the bound electron with energy W_0 is given by

$$\epsilon_b = 1 - W_0 \tag{10}$$

The kinetic energy of the free electron is given by

$$E = W - 1 \tag{11}$$

The transition energy ω is related to the latter two by

$$\Delta E = \omega = W - W_0 = E + \epsilon_b. \tag{12}$$

For a full subshell, the occupation probability has the value $\omega_\sigma = 2j + 1$. For a broken subshell ω_σ has a smaller value caused either by a smaller number of valence electrons or by punching holes into a subshell. For the neutral atom, of course

$$\sum_\sigma \omega_\sigma = Z. \tag{13}$$

The effect of internal conversion is accompanied by the significant X-ray radiation, which could positively have an effect on the use of scintillation detectors with the fine-dispersed gadolinium. We will only consider electrons with the energies higher than 20 keV.

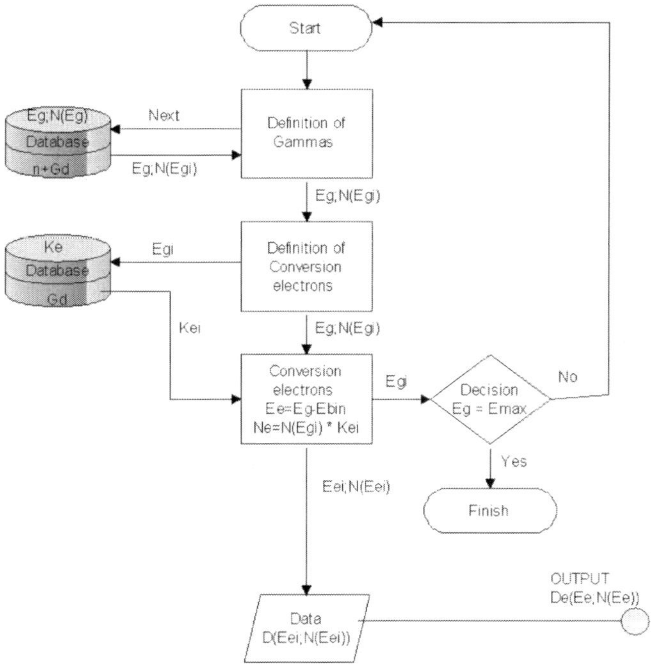

Figure 3. Diagram of algorithms for calculation of internal conversion electrons.

Data on coefficients of internal conversion are different in various sources [29, 30] that lead to the divergences in quantity of the secondary electrons. In our last modeling, we based on the last data presented in the database [31].

The low-energy gamma quanta are present at the spectrum and during their emission electrons from atomic subshells (so-called internal conversion electrons) are radiated with a high probability. The nuclear removes its excitation by radiating a gamma-quantum, but also a close located electron can be irradiated too. Usually K-electron (electron from K-subshell) is emitted, but also electrons from the higher subshells (like L, M, N and so on) can be emitted too. Vacancy of electrons (an electronic hole), formed as a result of this process, is filled by another electron from a higher level. This process is accompanied by radiation of X-ray quantum, or radiation of Auger-electron.

Table 4.

Electron Energies (keV)	Electron output 1/100 n	Electron path in Gadolinium (μm)	Energy of primary gamma quantum	Comment, level
29.3	35,58	4,7	79.51	K
34.9	13	6,29		K-Auger
71.7	5,57	20,7	79.51	L
78	1,2	23,78	79.51	M
131.7	6,96	55,70	181.93	K
174.1	0,99	86,27	181.93	L
180.4	0,21	91,23	181.93	M
205.4	0,14	111,47	255.66	K
227.3	0,16	130,27	277.54	K
729.9	0,03	649,38	780.14	K
893.85	0,06	830,05	944.09	K
911.8	0,04	849,83	960	K
926.8	0,03	866,35	975.4	K

In Figure 3 the scheme of algorithms for calculation of the internal conversion electrons are shown. First of all, it is necessary to define energies and quantity of gammas quanta's emitted in the reaction of radiation capture of neutrons by Gadolinium nuclear, Figure 4. This data can obtain from the data-base [31]. Energy of internal conversion electrons equal to energy of gammas minus of binding electrons energies of atomic subshells (K, L, M ...). For calculation of quantity of internal conversion electrons, it's necessary to define of internal conversion coefficients for Gadolinium for each atomic subshells.

The most intensive lines electrons are allocated in the table.

In the data in Table 4 the most intensive lines of electrons with the probability of emission higher than 0.03/100 neutrons, for the energies of primary gamma quantum less than 1 MeV. Data on Auger electrons which are formed during the K-shell filling are presented as well. In our calculations, totally 446 discrete electron energies with the output probability of more than 10^{-5} on 100 incidence neutrons were considered. In Figure 4 the histogram is presented, which shows the dependence of the most intensive lines of electron intensity on their energy.

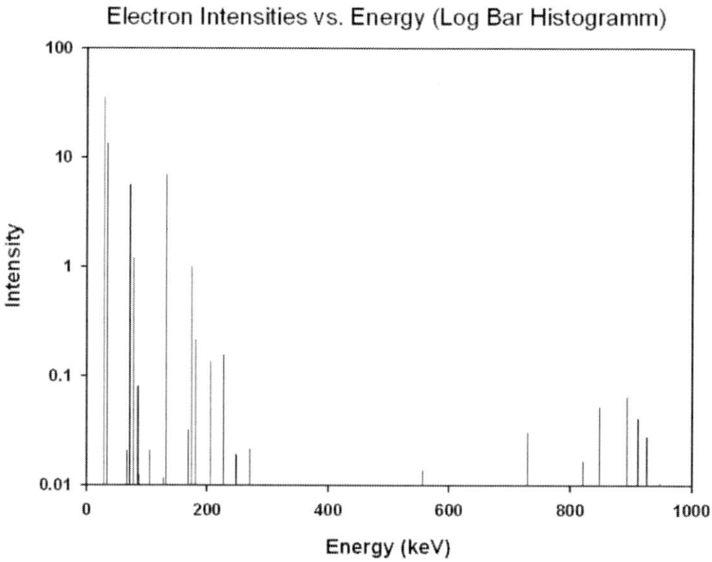

Figure 4. Intensity of internal conversion electrons emitted in the reaction ^{157}Gd (n,γ) ^{158}Gd depending on their energy (by 100 neutrons).

2.1.4. Intensity of Auger Electrons

Energy of emitted electrons defined by the energy of outcoming gamma quantum and the bound energy of electrons on the atom subshells

$$E_e = E_\gamma - E_{bi} \tag{14}$$

where, E_e is the energy of outcoming electron, E_γ – energy of gamma quantum, E_{bi} – bound energy of electrons on the atom shells.

Table 5.

E	K	L_1	L_2	L_3	M_1	M_2	M_3	M_4	M_5	N_1	N_2	O_1	O_2
$_{64}Gd$	50.24	8.38	7.93	7.24	1.89	1.69	1.55	1.22	1.19	0.38	0.27	0.04	0.03

In Table 5 the given bound energy of electrons on different gadolinium atom shells are shown [32].

By knocking out electrons, an electron hole (vacancy) is formed, which is filled by electrons from higher levels. During vacancy filling X-ray radiation, with the energy equal to the difference of bound energies at corresponding levels, is radiated

$$E_x = E_{bi} - E_{bj} \tag{15}$$

where E_x is the energy of X-ray radiation, E_{bi} and E_{bj} are bound energies of electrons at the corresponding atomic levels. Energy of excitation of atom can be removed also by the emission of Auger electrons. These electrons are emitted instead of X-ray quantum and possess energy equal to the energy of X-ray quantum minus bound energy of electron at the corresponding level

$$E_{ea} = E_x - E_{bi} \tag{16}$$

where E_{ea} – energy of Auger electron.

In Table 6 the most intensive X-ray lines and their relative outputs [33] are represented. Data are normalized to the intensity of line Ka1, whose intensity is chosen as 100%. Also, Augur electron data are represented, irradiation of them results in reduction of X-ray radiation output.

The effect of internal conversion is accompanied by the significant X-ray radiation, which could positively affect on the use of scintillation detectors with fine-dispersed gadolinium.

Probability of formation of Auger electrons have been derived from the theoretical emission probabilities through the relationship

$$I_{K-XY} = 100(1 - \omega_K)\frac{P_{K-XY}}{\sum P_{K-XY}}\ , \tag{17}$$

Table 6.

X-ray radiation			Auger electrons		
X-ray energy (average) (keV)	I_X (%)	Comment	Electron Energy (keV)	I_e (%)	Comment
6.06	42.6	L	4.84	201	L- Auger
42.31	55.6	Ka2 (KL2)	34.9	14.2	K- Auger
42.9962	100	Ka1 (KL3)			
48.7	30.8	Kb1 (KM(tot))			
50.0	8.9	Kb2 (KNO(tot))			

where $P_{K\text{-}XY}$ is the theoretical emission probability of a K-XY Auger electron, from Chen, et al. [33], ω_k is the K fluorescence yield, from Krause [34], and the summation is over all Auger electrons which are energetically possible.

Approximate Auger-electron energies can be calculated by the empirical Dillman quotation [35].

The average energy for a K-L_iX Auger transition is given by

$$\bar{E}(K - L_iX) = E_K - E_{L_i} - E_{M_3} - 0.75\left(E_{M_{3+}} - E_{M_3}\right). \quad (18)$$

and for higher atomic shells by

$$\bar{E}(K - XY) = E_K - 2E_{M_3} - 0.75\left(E_{M_{3+}} - E_{M_3}\right). \quad (19)$$

E_{li} is the bound energy of the L_i atomic shell for the element, E_K and E_{M3} are the corresponding bound energies of the K and M_3 atomic shells, E_{M3+} is the bound energy of the M_3 atomic subshell for the next higher element, and X and Y are designations for the higher atomic subshells. More precise Auger electron energies, one is referred to the publication of Larkins [36].

2.1.5. Passage of Electrons through a Substance

The probability of escape of electrons from a material of the converter can be considered on the basis of Bloch theory. For realization of calculations it is important to have exact data on brake ability of various substances for the charged particles. Brake ability is the average speed of loss of energy by charged particles in any point along their tracks.

For electrons and positrons full brake ability usually is divided on two components: a - the brake ability caused by collisions ("collision stopping power"), - average losses of energy on the unit of length of path due to non-elastic Coulomb collisions with the bound electrons of the matter, resulting in ionization and excitation; b - radiating brake ability ("radiative stopping power") - average losses of energy on the unit of length of path due to emission of brake radiation in the electric field of atomic nucleus and atomic electrons. Division of brake ability into two components is expedient for two reasons. First, methods of their determining are completely various. Second, the energy going on ionization and excitation of atoms, is absorbed by the matter rather close to the track of a particle, while the basic part of energy lost is the form of brake radiation, leaves far from a track before to be swallowed up. This distinction is important, when the attention is accented to the energy "transferred locally" to environment along a track, in distinction to the energy lost by an incident particle. Actually, the part of energy lost in the ionization impacts, turns to kinetic energy of secondary electrons and is transferred thus to some distance from a track of an initial particle.

In order to receive a rough estimation of locally transferred energy, it is expedient to enter the limited brake ability caused by collisions ("restricted collision stopping power") and to determine it as average losses of energy on the unit of length of the path due to events of excitation and ionization, in which the energy transmitted to secondary electrons, less than the chosen threshold.

Though brake abilities and path of electrons are widely used, they are seldom measured and turn out from the theory of energy losses. All data brought in tables contain the brake ability of electrons caused by collisions at the energies higher 10 keV, which is calculated according to Bethe theory (1930, 1932, 1933). Energy in 10 keV usually consider as the bottom limit of applicability of the theory. Basically, the non-trivial value describing properties of the substance in Bethe formula for brake ability is mean excitation energy, representing geometrical average of energies of excitation (atoms) of the substance, weighed in view of the appropriate oscillator forces. Except for the elements with very small nuclear number mean excitation energies are approximately equal to 10Z eV, where Z is the nuclear number. Exact calculations ab initio of mean excitation energies are possible now only for simple atomic gases. For the majority of substances, the mean excitation energy could be determined from experimental data.

Other important value in the formula for the brake ability, not contained in original Bethe theory, is correction on the density effect, reducing brake ability at polarization of substance by the relativistic charged particles [37]. Also at calculations it is necessary to take into account corrections on density effect by the Sternheimer method (1952) [38].

The linear brake ability caused by collisions with the dimension energy/length, is designated as — $(dE/dx)_{col}$ or S_{col}. It is frequently more convenient to consider the appropriate mass brake ability caused by collisions, $S_{col/\rho}$, where ρ is the substance density. Transition from the linear to the mass brake ability caused by collisions, substantially removes dependence on the density except for residual dependence due to the correction on density effect. If S_{col} is expressed in MeV • cm^{-1} and ρ – in g/cm^3 $S_{col/\rho}$ is expressed in MeV • cm^2• g^{-1}.

The brake ability caused by collisions is realized due to the energy transmitted by an incidence particle to connected nuclear electrons. We shall designate by *dσ/dW* differential cross section (on an atom electron) of non-elastic impacts with the energy transfer *W*. Then the mass brake ability caused by collisions is

$$\frac{1}{\rho}S_{col} = NZ\int W\frac{d\sigma}{dW}dW. \qquad (20)$$

where N is the number of atoms by 1 g of substance; Z is the nuclear number; $N = N_A/M_A = (uA)^{-1}$, where $N_A = 6{,}022045 \times 10^{23}$ mol^{-1} – Avogadro number; M_A is molar weight, g x mol^{-1}; A is the relative weight of atom (sometimes designated A_r) and $u = 1.6605655 \cdot 10^{-24}$ g is nuclear mass unit (1/12 weights of nuclide atom ^{12}C). Following formalism of Uehling (1954) [39], Bethe discuss results of estimation of the equation of brake ability [expression (20)]. These results are applicable to electrons and positrons, mesons, protons, α-particles and the heavy ions, completely deprived electrons. Energy W transferred to nuclear electrons in non-elastic impacts, divides on two classes depending on those, is W less or more than some threshold W_c which should satisfy to two conditions: a) W_c should be big in comparison with bond energy of nuclear electrons of the braking substance; b) the parameters of impact appropriate to losses of energy, smaller than W_c, should be big in comparison with the sizes of atoms.

Then the mass brake ability caused by collisions could be represented as the sum two components:

$$\frac{1}{\rho}S_{co1} = \frac{1}{\rho}S_{co1}(W < W_c) + \frac{1}{\rho}S_{co1}(W > W_c). \tag{21}$$

The basic result of Bethe theory with reference to electrons and the heavy charged particles could be expressed by the formula

$$\frac{1}{\rho}S_{co1}(W < W_c) = \frac{2\pi r_c^2 mc^2}{u}\frac{1}{\beta^2} \times \frac{Z}{A}z^2\left[ln\left(\frac{2mc^2\beta^2 W_c}{(1-\beta^2)I^2}\right) - \beta^2\right]. \tag{22}$$

where r_e is the classical radius of electron; mc^2 is the rest energy of electron; β is the speed of incidence particles in terms of speed of light; Z is an initial charge of a particle in terms of a charge of electron and I- is average energy of excitation of atoms of the substance.

Using numerical values Cohen and Taylor (1973) [40] for various physical constants, we receive, that

$$\frac{2\pi r_e^2 mc^2}{u} = -\frac{(2\pi)\cdot(7{,}940775\cdot10^{-26}\text{см}^2)\cdot(0{,}5110034\text{МэВ})}{1{,}6605655\cdot10^{-24}\text{г}} = 0{,}153536\text{МэВ}\cdot\text{см}^2\cdot\text{г}^{-1} \quad (23)$$

Expression (23) is fair when the speed of incident particles is greater than speed of nuclear electrons. With reference to electrons K-shell, it means that the condition $(Z/137\beta) < 1$ is satisfied.

Component of brake ability owing to close collisions it is estimated in approximation of free and bond electrons

$$\frac{1}{\rho}S_{co1}(W > W_c) = Nz\int_{W_c}^{W_m} W\frac{d\sigma}{dW}dW, \quad (24)$$

where $d\sigma/dW$ now is the differential cross-section of the energy transfer W in the collision with the free electron, and

$$W_m = \frac{2\tau(\tau+2)mc^2}{\left[1+2(\tau+1)\,{}^{m}\!/_{M}+\left({}^{m}\!/_{M}\right)^2\right]}, \quad (25)$$

is the greatest possible energy transfer; τ is the attitude of kinetic energy the incidence particles to their rest energy of rest; m/M is the ratio of weight of an electron to the weight of incidence particles.

For electrons the big transfers of energy to nuclear electrons, which are considered as free, are characterized by cross section (Moller (1932)) [41].

$$d\sigma = \frac{2\pi r_e^2 mc^2}{\beta^2}\frac{dW}{W^2}\left[1+\frac{W^2}{(T-W)^2}+\frac{\tau^2}{(\tau+1)^2(\left[{}^{W}\!/_{T}\right])^2}-\frac{2\tau+1}{(\tau+1)^2}\frac{W}{(T-W)}\right], \quad (26)$$

where $\tau = T/mc^2$ is the ratio of kinetic energy of an incidence electron to its rest energy. In the cross section Moller takes into account both relativistic and spin effects, and exchange effects due to in-distinguish-ability of incident electrons and electrons of a target. Conventionally the brake ability caused by collisions is calculated for fastest of the electrons formed in impact. The greatest possible transfer of energy W_m equal to T according to expression (2.37), in this case should be equal to $T/2$. Using in expressions (20), (21) and (22) the Miller section, we receive the following formula for the mass brake ability of electrons caused by collisions (Rohrlich and Carlson, 1953; Uehling, 1954г.) [42, 39]:

$$\frac{1}{\rho}S_{co1} = \frac{2\pi r_e^2 mc^2}{u}\frac{1}{\beta^2}\frac{Z}{A}\left[ln\left(T/_I\right)^2 + ln\left(1 + \tau/_2\right) + F^-(\tau) - \delta\right], \quad (27)$$

Where

$$F^-(\tau) = (1 - \beta^2)\left[1 + \tau^2/_8 - (2\tau + 1)ln2\right]. \quad (28)$$

Half of the value in the square brackets in the expression (2.39) represents the brake number on nuclear electron $L(\beta)$.

Corrections to Bethe theory, similar to corrections on shell-effect for brake ability, were discussed by Inokuti (1971) [43] concerning cross section of excitation. He marks, that these corrections contain the additional member proportional to the ratio m/M of electron mass to the mass of incident particles. Thus, one can expect, that in the case of electrons, for which $M = m$, corrections will be much more, than for protons. Probably the same happens concerning corrections of brake ability on shell-effect.

From measurements of brake ability $(S_{col/\rho})_{exp}$ it is possible to determine the mean excitation energy:

$$lnI = ln\left(\frac{2mc^2\beta^2}{1 - \beta^2}\right) - \beta^2 - x - \left[\frac{\left(S_{co1}/_\rho\right)_{exp}}{\frac{4\pi r_e^2 mc^2}{u}\frac{1}{\beta^2}\frac{Z}{A}z^2}\right]. \quad (29)$$

Where

$$x = {}^{C}/_{Z} + \frac{\delta}{2} - zL_1 - z^2L_2 \tag{30}$$

is the full correction uniting the corrections on shell-effect and the density effect, correction of Barkas and the Bloch correction.

Let ΔS_{col} is the error of the measured value S_{col}, and Δx is the error of a correction term x. We assume, that Δx and S_{col} are independent and could be incorporated quadratialy then the total error of estimated value I makes

$$\begin{aligned} \Delta I &= I[(\Delta L_1 \exp)\uparrow 2 + (\Delta x)\uparrow 2\,]\uparrow(1/2) = \\ &= I[(\Delta S_1 co1/S_1 co1\,)\uparrow 2\; L_1 \exp \uparrow 2 + (\Delta x)\uparrow 2\,]\uparrow(1/2) \end{aligned} \tag{31}$$

where L_{exp} is experimentally determined brake number [last term in the expression (31)], and ΔL_{exp} is the appropriate error. When in order to determine the values I one uses the data on run, the analysis of errors is more complicated and should take into account both the error of experimental value of run, and the error of the correction term x at all energies down to initial energy of incidence particles.

The full correction C on the shell effect is the sum of components C_K, C_L... for various nuclear shells. It is supposed, that the error owing to application of hydrogen-like wave functions is rather small for K-shell, is more essential to L-shell (especially for nuclear numbers Z < 30) and, probably, it is even more for the M - shell.

Extending of calculations on further shells is possible by application of more perfect wave functions, but could be very toilful. Bichsel [44] instead of this method applied semi-empirical numerical approach with the parameters determined from experimental data on the brake ability. He has accepted, that up to factor dependences C_M on speed of a particle shell effect are similar, and this assumption is extended also to farther shells. For L-shell with eight functions Walske [45] has received the function $C_L(\theta_L\ \eta_L)$, which depends on potential of ionization of L - shell through the parameter θ_L and on the energy of particle via the value

$$\eta_L = \left(\beta / \alpha Z^*\right)^2, \tag{32}$$

where $Z^* = Z — 4{,}15$ is effective nuclear charge for *L*-shell. The correction for the *M* – shell is calculated by the following ratio:

$$C_M = V_M C_L(\theta_L. H_M \eta_L). \tag{33}$$

where V_M = 1/8 of numbers of electrons on the *M*-shell, and H_M is adjustment parameter. The similar scale ratios were used for *N*-shell and for the combined *O*-*P*-shell.

In Figures 5 and 6 dependence of size of free path length electrons in gadolinium (g/sm2) and (μm) from their energy [46] is presented. Size *Re* can be defined by energy of electron and size specific ionization losses. The absorption coefficient F_0 characterizes probability of absorption of electrons in the substance. If *X* is the thickness of the converter, *Re* is the electron path in the material of the converter, then

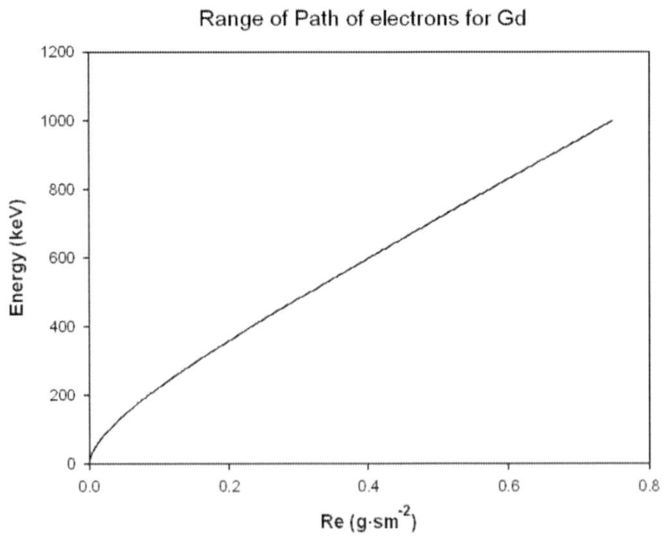

Figure 5. Dependence of the free path length of electrons on their energies in gadolinium (g/cm^2).

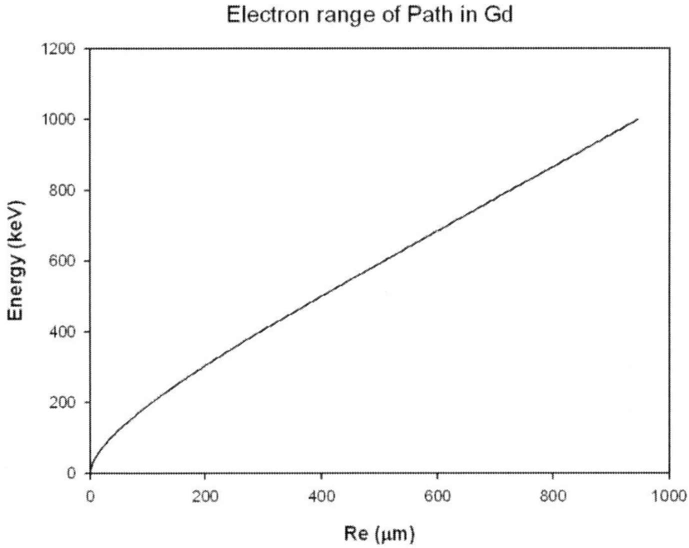

Figure 6. Dependence of the free path length of electrons on their energies in gadolinium (μm).

$$F_0(X) = 1 - X^* \rho / Re \tag{34}$$

where ρ is the density of the converter. For the gadolinium it is $\rho = 7{,}9$ g/cm^3.

Re is defined both by the electron energy and by the specific ionization loss value

$$\mathrm{Re} = \int_0^E dE(-dE / dX) \tag{35}$$

2.2. Model Representation and Calculation

Modeling was realized by examining the simplest targets, namely the plane-parallel foils. The calculations for thermal neutrons with the fixed energies, which correspond to the neutron wavelength of 1, 1.8, 3 and 4 Å were carried out; also both the thickness of converters from 1 μm to 40 μm, and isotopic composition of converter (for natural Gd and ^{157}Gd) were varied. In the calculation all electrons (appeared as a result of neutron capture act) are taken into account which are able to escape an infinite plane-parallel plate of the converter. The ratio of the number of electrons escaped from the foil to the number of incident neutrons is referred to as the efficiency of the converter.

Conventionally we divide the thickness of a foil into more thin components. For each elementary layer we count the probability of neutrons absorption with the fixed energy. Efficiency of the converter will be determined by the sum of probabilities of neutron absorption and probability of the electron escape from body of converter. In order to calculate the electron escape of converter we have chosen a simple model, namely, geometrical. The choice of the model is made from the following assumptions: all electron emissions are isotropic, the length of path for any fixed energy of electrons (Re_i) is constant (fluctuation of power losses in the end of path is neglected). Then the density of probability to find electrons in the material forms a sphere with the radius equal to Re_i (for each fixed energy). If the center of the sphere is crossed by the plane, two identical hemispheres are formed, which correspond to the electron escaping to the forward and backward hemisphere, thus areas of hemispheres could be considered as the probability of an electron output. In this case total probability is 100%, and escape to the one of hemispheres is 50%. If we begin to cross a sphere with a step much less than Re_i, segments will be formed whose area will be equal to probability of the electron escape. The step of iterations should be at least 100 times less than Re_i, and then the electron absorption under the big angles could be neglected. The area of a segment and accordingly probability of the electron escape becomes equal to zero in the intersection of a sphere by the plane at the distance Re_i. The sum of probabilities of electron escapes for all energies electrons, taking into account their weight contribution will determine the total electron escape probability.

The scheme of algorithms calculation of neutron absorption in a converter material is shown in Figure 7. In the beginning the select of a material of the converter (natural Gadolinium or its 157 isotope), and a neutron wavelength is necessary. In calculations we are use four fixed neutron energies; these are conforming for neutrons with wavelengths of 1, 1.8, 3 and 4 Å. After a select of parameters from a database data about cross section of neutron interaction with materials [20] retrieved. Calculation of neutron absorption is made for each selected layers.

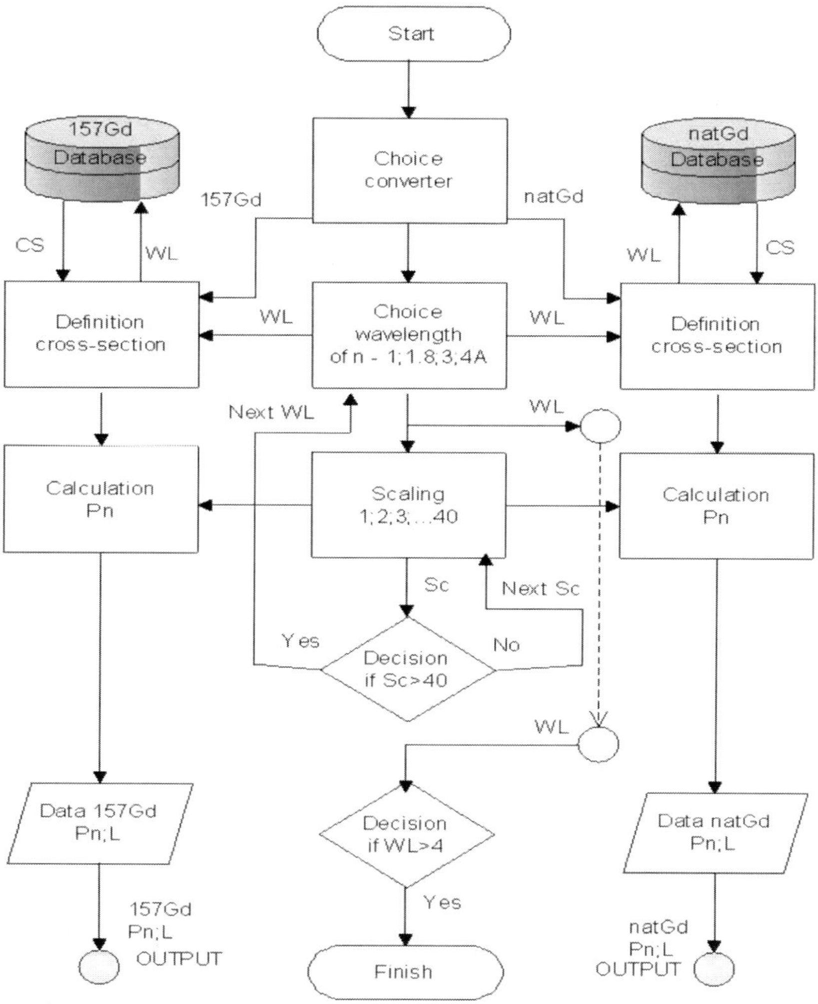

Figure 7. Calculation algorithm diagram of absorption path of neutrons in substance.

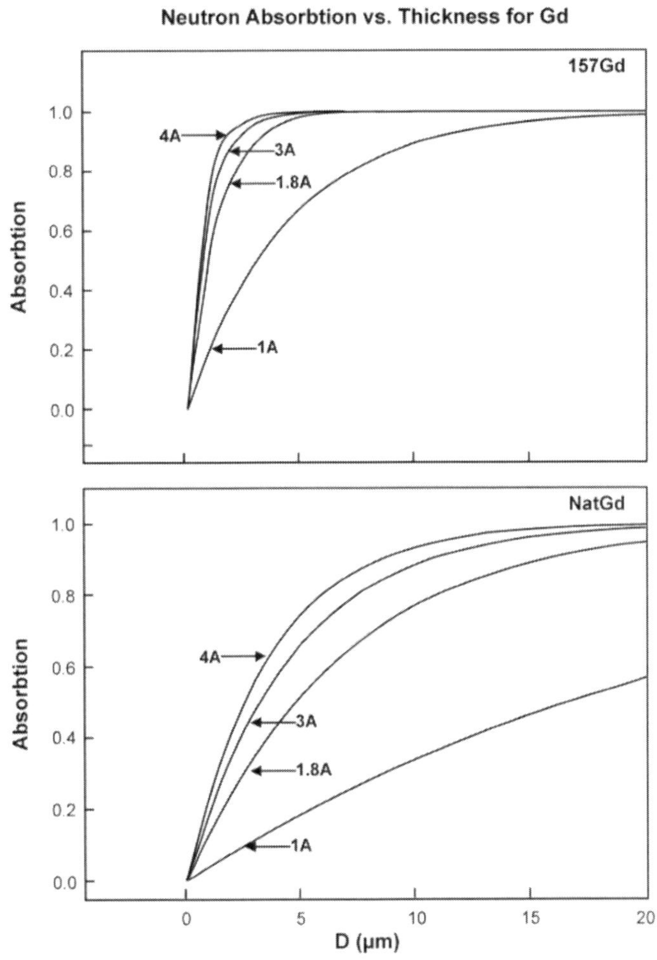

Figure 8. Curve uptakes of neutrons for lengths of waves 1, 1.8, 3 and 4 ÅÅ for natural gadolinium and it 157 isotope.

Results of calculation of neutron absorption in natural Gadolinium and its 157 isotope are shown in Figure 8. Figure show that at 30 microns thickness of the natural Gadolinium converter the neutrons with the wave length above 1.8 Å are absorbed completely. For its 157 isotope for neutrons with the wave lengths >1.8 Å the same absorption happens at 8 microns thickness of the converter.

Probability of electrons escape from a converter can be considered by the Bethe- Bloch theory. For realization of calculations it is important to have exact data on brake ability of various substances for the charged particles. Brake ability is average speed of lose of energy by charged particles in any point along their tracks.

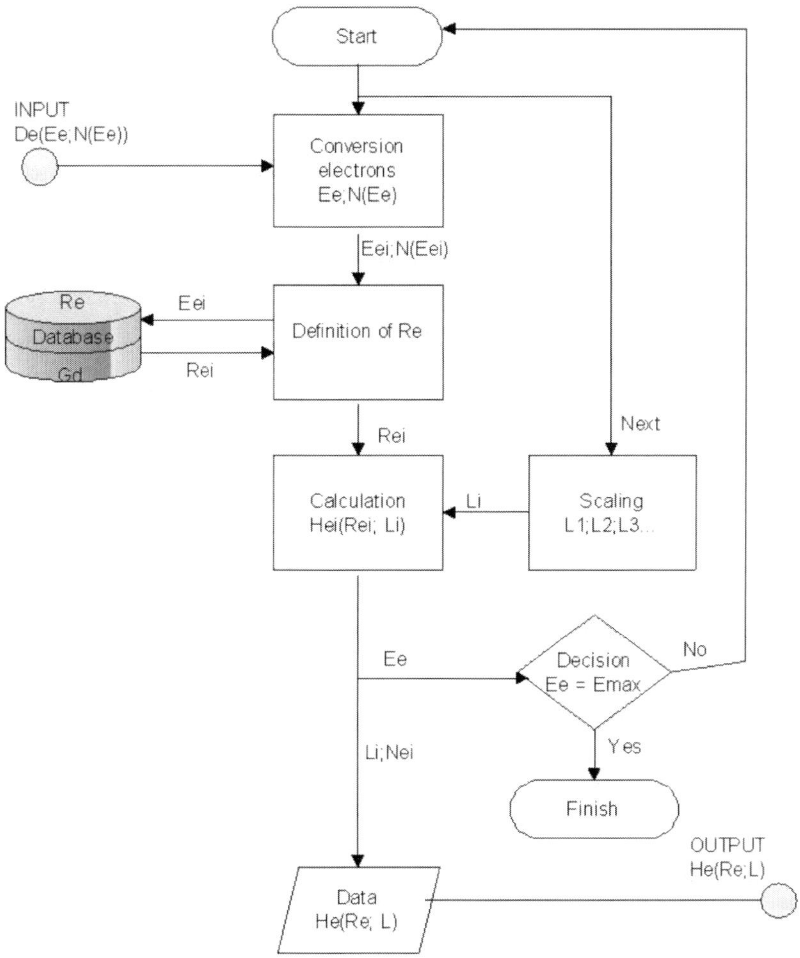

Figure 9. Scheme of algorithm calculation of ability an electrons exit from the converter body.

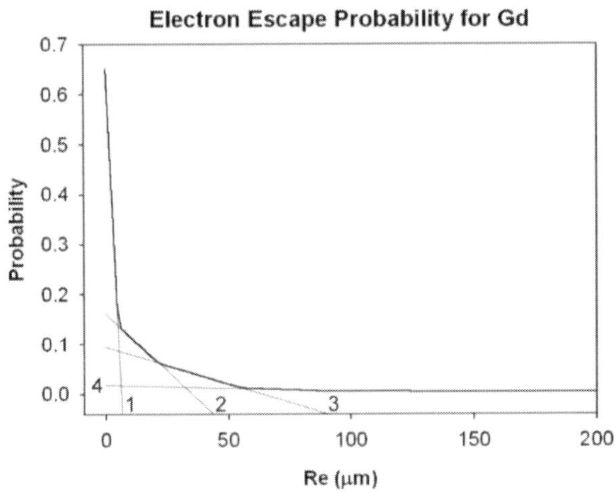

Figure 10. Probability of yield of isotropic emitting secondary electrons from gadolinium foils for various thickness, in view of their intensity.

Calculations of exit probability of isotropic emitting electrons, originated in the ^{157}Gd (n, γ) ^{158}Gd radiation capture reaction, are conducted. In Figure 6 one can see contributions of electrons of different energetic groups. Calculations were made for conditionally fixed point of conversion of neutrons having coordinate (X = 0, Y = 0).

The scheme of algorithm calculation of ability an electrons exit from the converter is shown in Figure 9. For calculations of probabilities of electrons output from substance the database about internal conversion and Auger (Figure 4) downloaded. For each discrete energy of an electron the database [46] about of free length (Re) electron in substance were used. For each elemental stratum probability of an exit of electrons is yielded, for calculation the data on Re and intensity of electrons used.

Separated energies of electrons could be divided into 4 groups (Figure 4). In the first group the least energetic and therefore having small ranges in the substance, and in the fourth one the most energetic electrons are placed. In Figure 10 one can see contributions of electrons of different energetic groups. Calculations were made for conditionally fixed point of conversion of neutrons having coordinate (X = 0, Y = 0). The contribution of 4 groups of electrons with different energies is well seen.

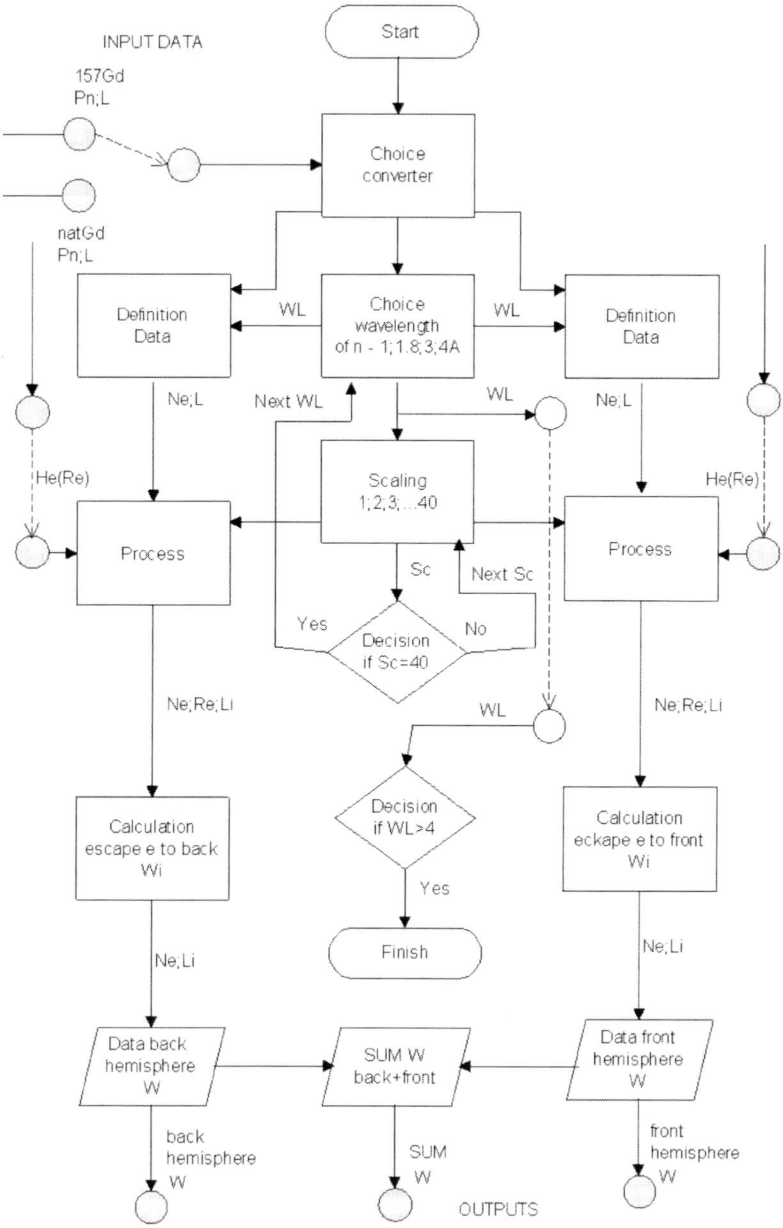

Figure 11. Scheme algorithm calculation of the efficiency of converters.

In calculations, the probability of neutron absorption for each elementary layer, as a result of reaction of inelastic interaction (n,γ), is determined, using database for fixed neutron energies. The probability of electron emission and their escape probability from the body of the converter were calculated.

Calculations of converters efficiency have been executed for the foils made from natural Gadolinium and it 157 isotope, for four fixed wavelengths of neutrons 1, 1.8, 3 and 4 Å. Calculation were performs for geometry of an exit of electrons in a back, forward hemisphere and their sum for different thickness of the converter.

The scheme of algorithms of calculation is given in Figure 11. In the beginning the select of the type of converter and a neutron wavelength was made. The thickness of the converter is divided into thin elemental stratums. For each stratum the corresponding data on absorption of neutrons (Pn, L) download. During calculation to each stratum conventionally coordinates X0, Y0 appropriated, the data on probability of an exit of electrons (He(Re)) is download. The data gained for each stratum tots. At modeling efficiency calculations separately for geometry of an exit of electrons in a forward and back hemisphere and their sum were made.

Obtained data on efficiency and optimal thicknesses of converters for natural gadolinium and its 157 isotope; shown in Table 7 and Table 8 correspondingly.

Table 7. Calculated data on efficiency and optimal thicknesses for natural gadolinium converter foils

λ(Å)	Efficiency E and optimal thickness T		(μm)
	Forward	Backward	Total
1	0.044 (18)	0.064 (40)	0.10 (24)
1.8	0.096 (5)	0.135 (30)	0.208 (7)
3	0.122 (4)	0.164 (25)	0.261 (5)
4	0.138 (4)	0.182 (20)	0.295 (4)

Table 8. Calculated data on efficiency and optimal thicknesses for ^{157}Gd converter foils

λ(Å)	Efficiency E and optimal thickness		T (μm)
	Forward	Backward	Total
1	0.126 (4)	0.169 (18)	0.270 (5)
1.8	0.208 (3)	0.258 (12)	0.448 (4)
3	0.232 (2)	0.277 (7)	0.489 (3)
4	0.241 (2)	0.287 (5)	0.516 (2)

Results of calculations of neutron absorption in converters made from natural gadolinium and its 157 isotope are shown in the Figure 8. Figure shows that at 30 microns thickness of the natural gadolinium converter the neutrons with the wave length above 1.8 Å are absorbed completely. For its 157 isotope for neutrons with the wavelengths >1.8 Å the same absorption happens at 8 microns thickness of the converter.

In order to detect thermal neutrons by thc ^{157}Gd converter we could limit ourselves by the thickness of 5 microns of the converter, if there no technological restrictions. It should be taken into account that the majority of electrons emitted in the reaction of radiating capture of neutrons have ranges less than 5 microns. When using natural gadolinium the situation is more complex, since low absorption requires converter to be thicker than 20-40 micron.

Figure 12 and Figure 13 show the dependence of registration efficiency on converter thickness for neutrons with the different energies.

Obtained results of calculations are compared with the experimental data presented in the paper [14], which data were received in the reactors of Atominstitut in Vienna (ATI) and the ILL Grenoble.

In this paper experimental data of the detection efficiency was measured in backward direction for six different energies and compared with a calibrated ^{3}He counter. In this work converter made from natural gadolinium and enriched up to 90.5% ^{157}Gd were used.

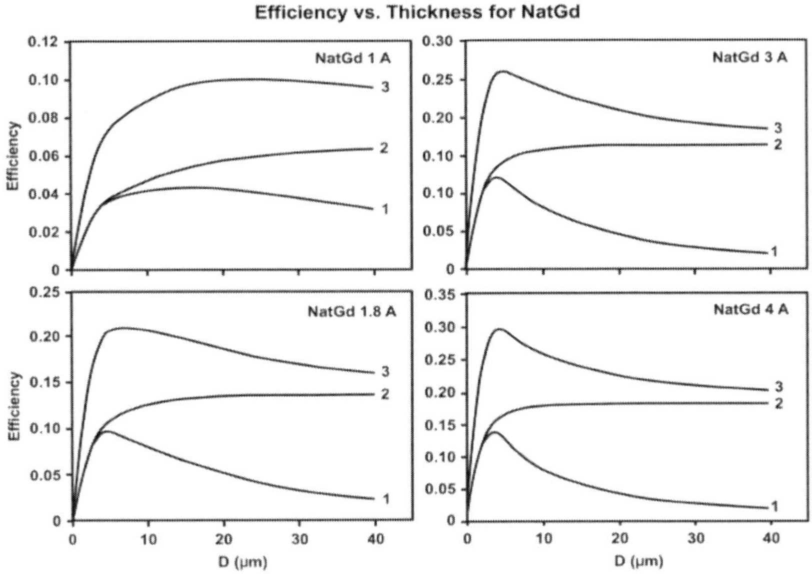

Figure 12. The dependence of the neutron registration efficiency of the natural gadolinium converter on its thickness. Curve 1 and 2 correspond to emission of electrons into the front and back hemisphere, respectively, curve 3 is their sum.

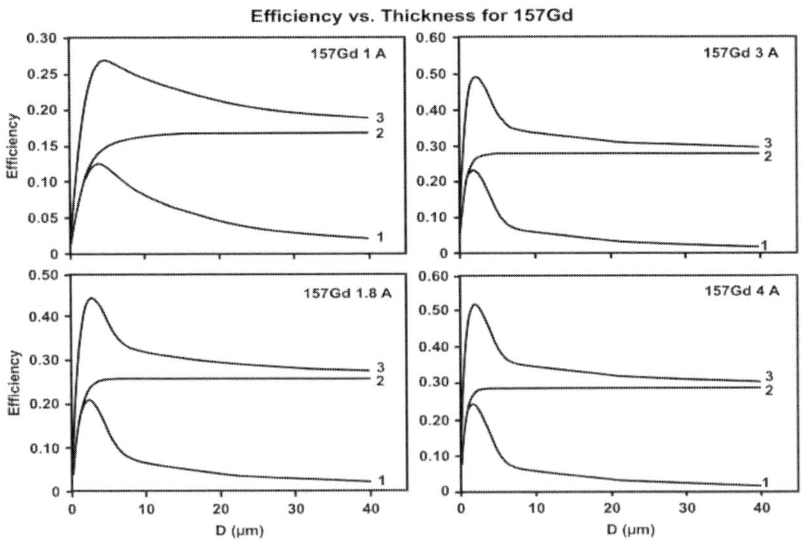

Figure 13. The dependence of the neutron registration efficiency of the 157 isotope of gadolinium converter on its thickness. Curve 1 and 2 correspond to emission of electrons into the front and back hemisphere, respectively, curve 3 is their sum.

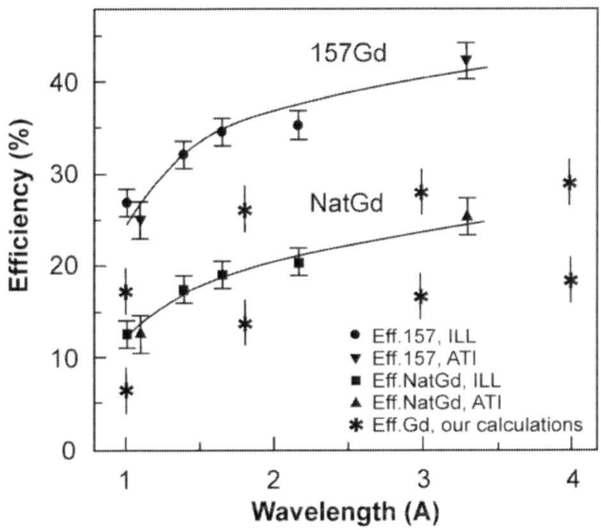

Figure 14. Comparison of our calculated data with the experimental data for backward escape geometry. Experimental data are obtained in Atominstitut in Vienna (ATI) and the ILL Grenoble [14].

The effect of comparison is shown in Figure 14 for natural gadolinium a curve of calculations lies a little below experimental data. Errors of calculations are caused both by the precision of determining of gamma-quanta output and by determining of neutron cross-section. For the converter made from 157 isotope the experimental curve is little bit higher (at the account of electrons with energy more than 29 keV) that testifies to necessity of the account of more thin effects.

2.2.1. Probability of Converters Activation

Deficiency of detectors with natural gadolinium converters is possible activation of converters, which is accompanied by radiation of retarding gamma quanta and electrons which, in turn can form a spots with increased background level. The basic contribution is carried out by long-lived isotopes (Table 2).

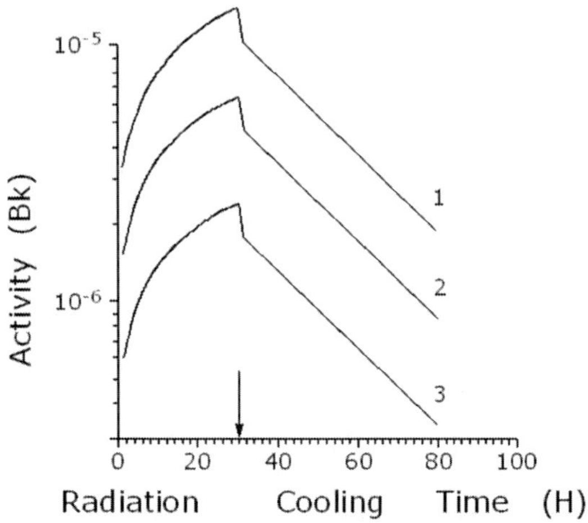

Figure 15. The background level of converters from natural gadolinium with thickness 1, 3 and 10 microns (accordingly curves 1, 2 and 3) at exposure on a bundle of neutrons intensity 1n/sm^2 *s with irradiation time of 30 hours [9].

The activity $A(t)$ (particles s^{-1}) at a time t seconds after the removal of a thin foil from a neutron flux having an energy spectrum $\Phi(E)$ in which the foil has been irradiated for T seconds is given by expression [47]

$$A(t) = (V \rho\ Na/A)\ (1\text{-}exp(-\lambda T)\ exp(-\lambda T) \int \sigma_{act}\ (E)\ \Phi(E)\ dE, \qquad (36)$$

where V, ρ and A are the volume, density and atomic weight of the foil material respectively, N_A is Avogadro's constant, λ (s^{-1}) is the decay constant of the radioactive species produced by the irradiation and $\sigma_{act}(E)$ is the activation cross-section of the material. Values of σ_{act} averaged over the thermal and fission energy regions are given in the table together with the activation resonance integral.

Calculations of possible activation of converters made from natural gadolinium have been fulfilled, and a background (noise level) which will accompany decay of the activated isotopes. At calculations of level of the directed background emitted electrons have been considered only. In Figure 15 effects of calculations for foils with a thickness 1, 3 and 10 microns (curves 1, 2 and 3) are presented. Calculations have been satisfied for a requirement that quantity of a neutron stream makes 1n/sm^2 *s and irradiation time of 30 hours. In it cases maximum noise 3 microns of converters occurring as a result of activation will not exceed 4 x 10^{-5} Bk on an individual neutron. In too time trace of activation can be considered at post processing of the data analysis.

2.2.2. Modeling for a Case of Neutrons Flow to Converter under Various Angles

In the process of calculation, we varied the angle of incidence of neutrons flow onto the foil. Under the small angle of fall we can increase neutron path in the material of the converter, yield path for the output of the secondary electrons will be small and constant. The most ideal case would be a direction of a bunch of neutrons along the converter of small thickness. This case cannot be realized in the practice, but yet it is possible to estimate theoretical for calculation of maximum possible efficiency.

On the Figure 16 the dependence of efficiency on the incidence angle of neutrons for ^{157}Gd for two various thicknesses of converters (1 and 3 microns) are shown. From figure it can be seen, that at small thickness of the converter (1 micron and less) difference of electron outputs in frontward and backward of hemisphere is not so significant. The probability of an output of electrons makes approximately 32 and 35% respectively for an output of electrons in frontward and backward of hemispheres. Total efficiency reaches 64% [9, 48], which can be considered as the greatest possible efficiency for gadolinium converters. The difference becomes more substantial when thickness of converters is increased.

Similar calculations are made as well for converters from ^{nat}Gd. From Figure 17 it is clear that at extremely small angles of neutron incidence (2-3°), converters from natural gadolinium can compete with the converters from ^{157}Gd.

The calculations made for a case of neutron incidence under angle of 10° for various thicknesses of converters for length of a wave of neutrons 1,8Å, for two types of converters natural gadolinium and its 157 isotope. Using converters from ^{157}Gd, at a thickness of 1 micron it is possible to reach efficiency of 64% which is a theoretical limit.

Obtained as a result of modeling calculations data, allows looking with optimism at possibility of making of effective position-sensitive detectors of thermal neutrons with solid-state converters from the natural gadolinium which efficiency of registration becomes comparable with efficiency of detectors filled ^{3}He, in too time these detectors will be much more low-cost and are more easy-to-work.

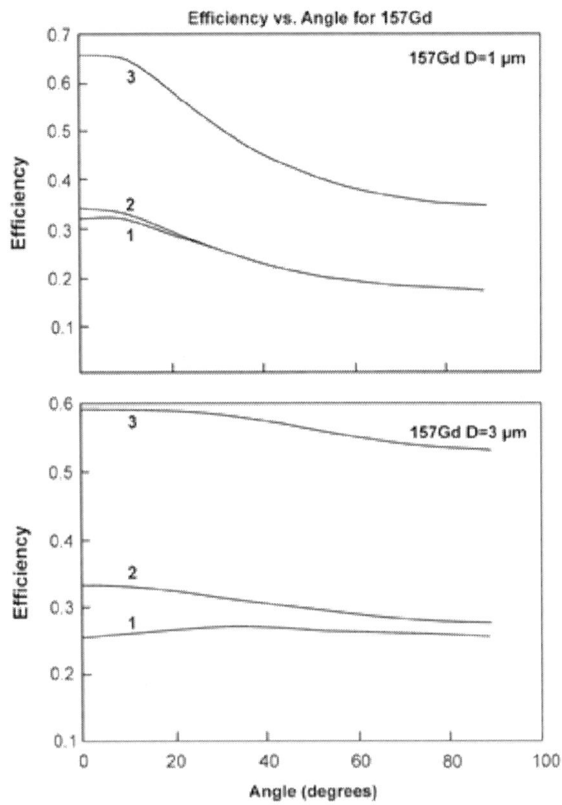

Figure 16. Dependence of efficiency of registration of neutrons with a wavelength 1.8 Å from an angle of neutrons failing, for ^{157}Gd for thickness of converters 1 and 3 microns. A curve 1 and 2 accordingly an escaping of electrons in forward and back hemispheres, the curve 3 their sum.

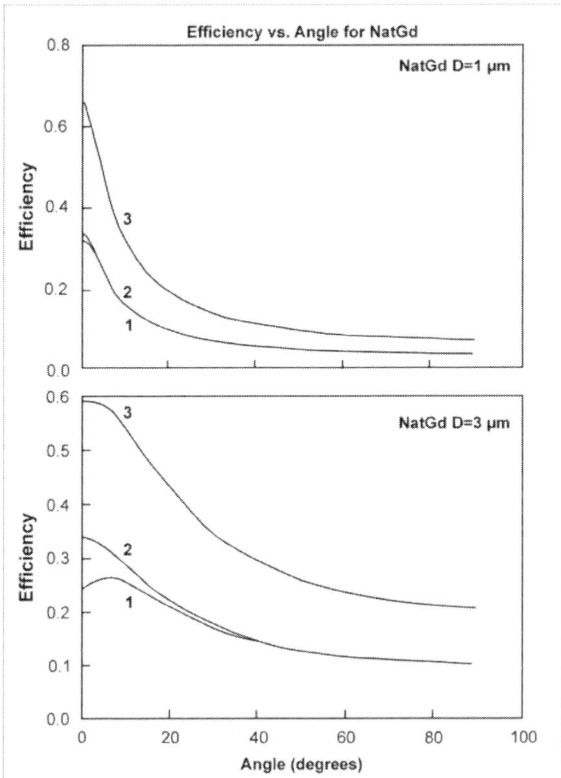

Figure 17. Dependence of efficiency of registration of neutrons with a wavelength 1.8 Å from an angle of neutrons failing, for ^{nat}Gd for thickness of converters 1 and 3 microns. A curve 1 and 2 accordingly an escaping of electrons in forward and back hemispheres, the curve 3 their sum.

For example, on Figure 19 the simplified plan of the one-coordinate detector of thermal neutrons on the basis of two multistep avalanche chambers with a natural gadolinium converter is presented. The detector will be oriented at an angle 10^0 to an axis of slope of a neutron beam. In Table 9 detector key parameters [9] are given.

The detector will possess small sensitivity to gamma – irradiation and high counting rate which will be depend by a used method of removal of the information. Efficiency of registration of the detector will be comparable with detectors on a base of ^3He convertor at much best space resolution.

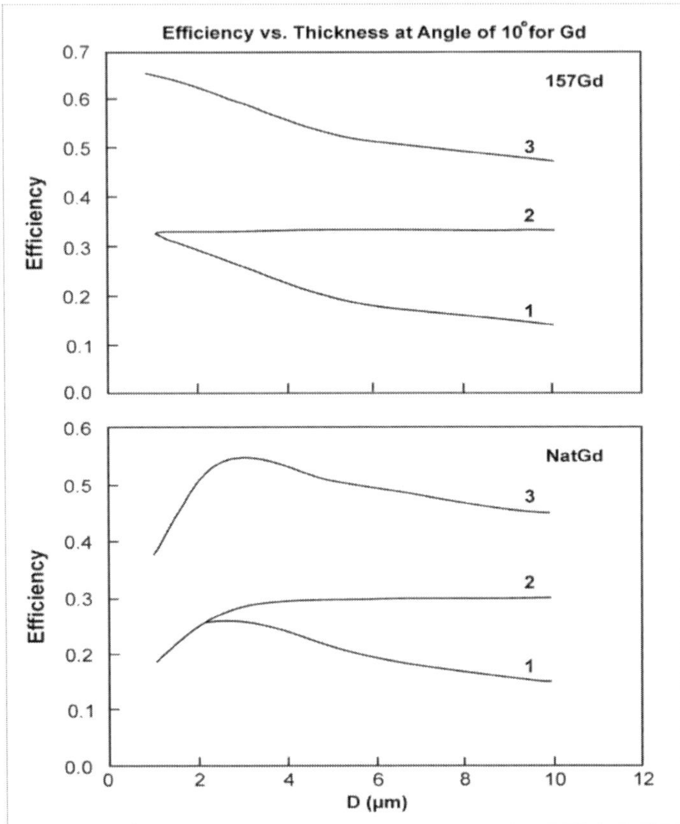

Figure 18. Dependence of efficiency of registration of thermal neutrons with a wave length 1,8 Å from a thickness of the converter at an angle 10^0 of flowing neutrons.

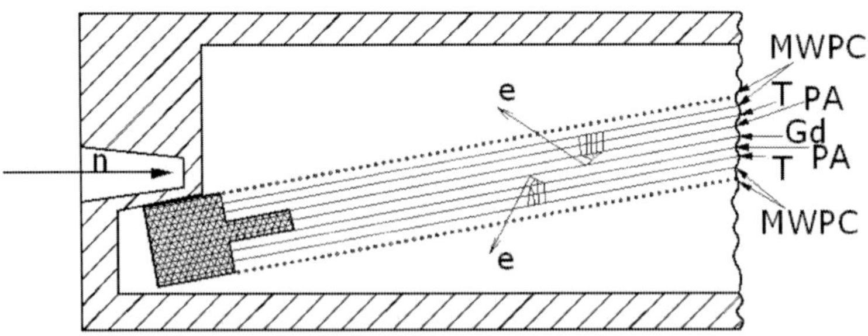

Figure 19. Schematic view of the detector, where MWPC- multiwire proportional chamber, T-transfer gap, PA- preamplification gap, Gd-gadolinium converter.

Table 9.

Sensitive area	(200-350) x 10 мм^2
Efficiency of registrations	
λ=1 Å	35%
λ = 1.8 Å	55%
λ = 3 Å	58%
λ = 4 Å	59%
Spatial resolution	0.8 мм
Time resolutions (on anode)	1-3 x10^{-9} c.
Gas mixture	Izo-butane
Producing pressure	5-20 torr

2.2.3. Contribution of Low-Energetic Electrons to General Efficiency of Converters

In our carlier calculations we have taken into account only electrons, which have energy higher than 29 keV. These energies of electrons have been chosen not only in our calculations, but also by other authors. Thus not taken into account the low-energetic Auger electrons, i.e., Auger electrons from the L-subshell with the energy of 4.84 keV and Auger electrons from the M-subshell with the energy of 0.97 keV. This electrons have rather small free path length in gadolinium: 0.3 microns (4.84 keV) and 0.04 microns (0.97 keV). They bring a small contribution to the general efficiency at use of converters made from natural gadolinium as the free path length of neutrons in natural gadolinium makes tens micron. At the same time their contribution becomes rather essential at use of converters made from 157 isotope of gadolinium as the free path length of neutrons in them does not exceed 2-3 microns and this length becomes comparable with the length of free path length of electrons.

In the literature and databases there is none data on free path length of electrons with energy less than 10 keV in various materials. All data begins with this energy. All tabular data is made on the basis of the theory Bette - Bloch and energy 10 keV is lower limit of usability of the yielded theory.

For calculation of free path length of electrons of Auger in gadoliniums we had to extrapolate available data to almost zero-point energy. Effects of extrapolation are given in Figure 20. From figure it is visible that the maximum free path length of electrons with energy 4,84 keV can makes 0,3 microns, and for electrons with energy 0,97 keV only 0,04 microns. Similar free path length is rather small, especially in comparison with run of neutrons in natural gadolinium. In too time for converters from 157 isotope of gadolinium run in 0,3 microns makes more than 10% from a free path length of neutrons and it can increase efficiency of converters essentially.

There are different data on the quantity of Auger electrons in the literature, so the quantity of electrons from the K-subshell makes from 10 up to 14% [23, 32], these data are normalized to the intensity of X-ray radiation Ka1 (KL3) which intensity is chosen for 100%. The data on the quantity of electrons from the L-subshell even more differ, from 150 up to 200 [23, 32]. There are no data for electrons from the M-subshell at all. At the choice of quantity of electrons from the M-subshell we started with the assumption that the quantity of Auger electrons grows at distance from a nucleus. So the quantity of electrons grows approximately 15 times at the transition from the K-subshell to L-subshell.

This tendency is kept further too, i.e., from L to M-subshell and further, N and O. At the Auger effect the external electron subshells are peeled from the electrons practically completely. All data on the quantity of electrons are usually normalized to Ka1 (KL3) of X-ray line, which is the most intensive one. At radiating capture of neutrons by gadolinium nuclei radiations of internal conversion electrons occurs. The probability of emission of internal conversion electrons makes approximately 60%. Auger electrons is accompanying emission of internal conversion electrons. Radiation of internal conversion electrons occurs their vacancies on electron shells filled by the electrons from the higher shells. This process is accompanied by X-ray radiation and Auger electrons. In our calculations, the quantity of Auger electrons it is necessary to normalize not to Ka1 (KL3) X-ray line, but to the quantity of internal conversion electrons, i.e., the quantity of electrons depends on the factor of internal conversion.

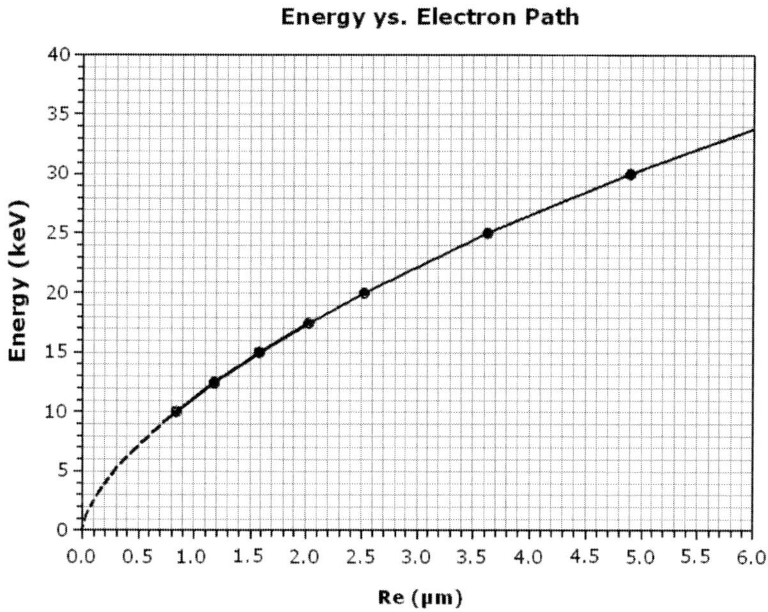

Figure 20. Curve of dependence of electron runs in gadolinium depending on their energy, for a low-energy range.

In Table 9 the most intensive lines of electrons emitted during the process of radiating capture of thermal neutrons by gadolinium nuclei are presented. The data on Auger electrons L-Auger and M-Auger are added to the Table 9.

Converters made from 157 isotope of gadolinium have an abnormal high cross-section of interaction with thermal neutrons so the cross-section makes 253778.40 barns for neutrons with the wavelength of 1.8 Å. The cross-section strongly grows with the increase of neutron wavelength. Converters with the thickness of 2.5 microns absorb more than 80% of neutrons with the wavelength of 1.8 Å and more than 90% of neutrons with the wavelengths more than 3Å, see Figure 10.

Another situation takes place at use of converters made from natural gadolinium. So the cross section of interaction makes 48149.41 barns for neutrons with the length of wave 1.8 Å. 80% of attenuation of a neutron beam (1.8Å) happens at the thickness more than 12 microns.

Table 10.

Electron energy (keV)	Electron output 1/100 n (error)	Electron path in gadolinium-microns	Energy of initial gamma-quantum	Comments, level
0.97	>200	0.04		M-Auger
4.84	97(33)	0.3		L-Auger
29.3	35.58	4.7	79.51	K
34.9	7.9(4)	6.29		K- Auger
71.7	5.57	20.7	79.51	L
78	1.2	23.78	79.51	M
131.7	6.96	55.70	181.93	K
174.1	0.99	86.27	181.93	L
180.4	0,21	91.23	181.93	M
205.4	0,14	111,47	255.66	K
227.3	0,16	130,27	277.54	K
729.9	0,03	649,38	780.14	K
893.85	0,06	830,05	944.09	K
911.8	0,04	849,83	960	K
926.8	0,03	866,35	975.4	K

The analysis of curves shows, that at use 157 isotope of gadolinium, the small run of Auger electrons (<0.3 microns) can increase the general efficiency of converters essentially.

Calculations of the efficiency of gadolinium foils are carried out, for the natural gadolinium and its 157 isotope, for four fixed wavelengths of neutrons depending on the thickness of converters. The figures, 12 and 13 show the results of calculations without taking into account of the influence of low-energy Auger electrons. The figures, 21 and 22 show the results of similar calculations, but here the low-energy electrons taken into account.

One can see from figures that with the increase of neutron wavelength the difference in the efficiencies is increased. Especially is well visible at use 157 isotope. So for neutrons with the wavelength of 4Å the total efficiency grows practically by 10%. Comparison of result of calculations for a case of recording of electrons with energy is more 29 keV and more than 0,93 keV are given In Table 11.

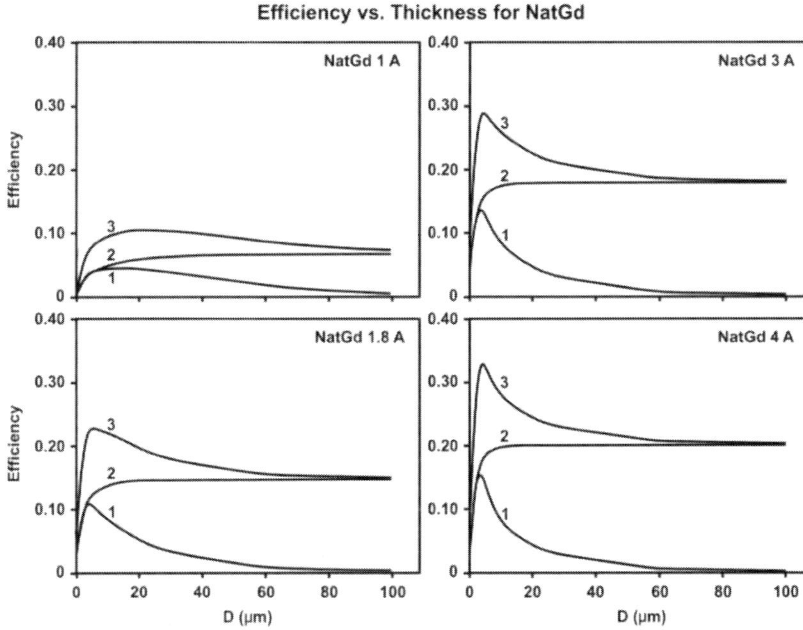

Figure 21. Curves describing the efficiency of converters made from natural gadolinium depending on the thickness of the converter, taking into account the low-energy Auger electrons. The curve 1 characterizes electron emission in a direct, the curve 2 - in back hemispheres. The curve 3 is their sum.

Table 11.

Wave-length (Å)	Forward	Backward	Total	Wave length (A)	Forward	Backward	Total
	^{Nat}Gd (E> 29 keV)				^{157}Gd (E> 29 keV)		
1	0,0435	0,0635	0,107	1	0,1262	0,1686	0,2948
1,8	0,0967	0,1357	0,2324	1,8	0,2098	0,258	0,4678
3	0,1218	0,1636	0,2854	3	0,2319	0,2772	0,5091
4	0,1379	0,1816	0,3195	4	0,2443	0,2865	0,5308
	^{Nat}Gd (E> 0,93 keV)				^{157}Gd (E> 0,93 keV)		
1	0,0462	0,068	0,114	1	0,1402	0,1851	0,3253
1,8	0,1063	0,1464	0,2527	1,8	0,2342	0,3103	0,5445
3	0,1335	0,1792	0,3127	3	0,2633	0,3501	0,6134
4	0,1509	0,2011	0,352	4	0,2848	0,3756	0,63

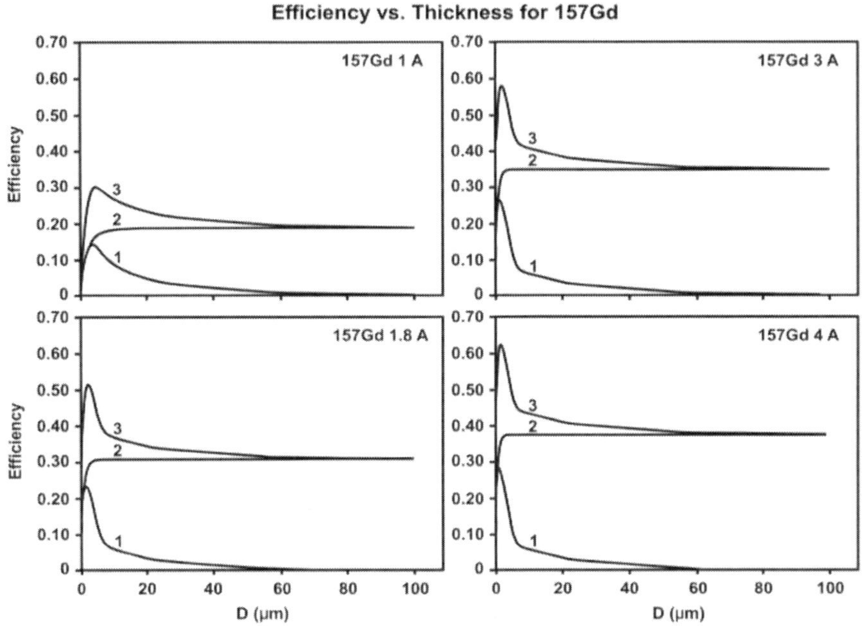

Figure 22. Curves describing the efficiency of converters made from 157 isotope of gadolinium depending on the thickness of the converter, taking into account the low-energy Auger electrons. The curve 1 characterizes electron emission in a lobby, a curve 2 - in back hemispheres. The curve 3 is their sum.

Received our data were compared with the experimental data given in the paper [14]. In this work the experimental data on the efficiency of detecting of neutrons emitting in a back hemisphere for 6 various energies and their comparisons with the calibrated ^{3}He counter are shown. In this work converters made from natural gadolinium and enriched up to 90,5% ^{157}Gd were used. The work was carried out on the reactors of Atominstitut in Vienna (ATI) and the ILL Grenoble (Figure 23).

One can see from the figure that in the case of the account of the contribution of low-energy Auger electrons (electrons with the energy > 0.93 keV) the results of our calculations well coincide with the experimental data. This concurrence is well visible for the converters made from 157 isotope of gadolinium. If we do not take into account the low-energy electrons the curve lays much below experimental data.

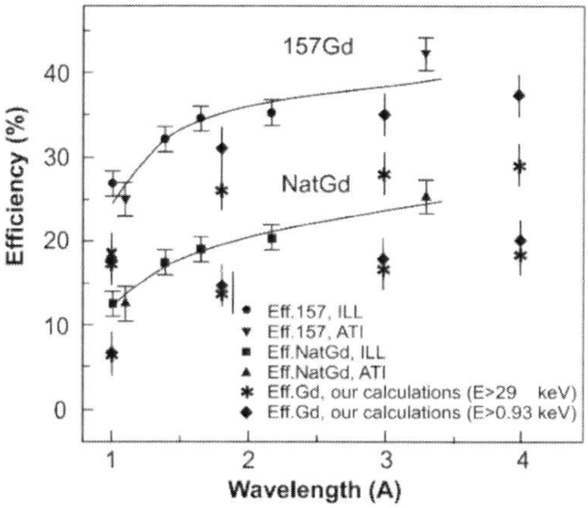

Figure 23. Comparison of the results of our calculations with the experimental data presented in the paper [14]. Calculations are conducted for two boundary energies of taken into account electrons; higher than 29 keV and higher than 0,93 keV.

2.2.4. Contribution X-Rays and Soft Gamma Radiations on General Efficiency of Converters

For correct calculations of the efficiency of converters it is necessary to take into account all secondary electrons. Earlier we took into account only internal conversion electrons and Auger electrons. But in the process also secondary electrons appearing during absorption of X-ray radiation and gamma-quanta can participate. These are photoelectrons, Compton electrons, electron-positron pairs. During radiation of internal conversion electrons vacancies on electron shells which are filled by electrons from higher shells occur. At filling of vacancy X-ray quanta or Auger electrons are radiated.

We have estimated the contribution of X-ray and low-energy gamma-quanta absorbed directly in the material of the converter and resulting in occurrence of secondary electrons. At practical calculations absorption of quanta in materials of detectors also should be taken into account. At such calculation total efficiency can appear higher, as the part of electrons can be formed from quanta in a material of the detector. In the calculations we take into account only an escape of electrons from a material of the converter.

Internal conversion coefficient is defined as follows:

$$\alpha_{ik} = \frac{I_\varepsilon}{I_\gamma}, \tag{37}$$

where I_ε is the intensity of conversion electron, and I_γ is the intensity of gamma radiation. I_γ is well known (Table 3).

Energy of emitted electrons defined by energy of outcoming gamma quantum and bound energy of electrons on the atom subshells

$$E_e = E_\gamma - E_{bi} \tag{38}$$

where, E_e is the energy of outcoming electron, E_γ – energy of gamma quantum, E_{bi} – bound energy of electrons on the atom subshells.

By knocking out electrons, an electron hole (vacancy) is formed, which is filled by electrons from higher levels. During vacancy filling X-ray radiation, with the energy equal to the difference of bound energies at corresponding levels, is radiated

$$E_x = E_{bi} - E_{bj} \tag{39}$$

where E_x is the energy of X-ray radiation, E_{bi} and E_{bj} are bound energies of electrons at the corresponding atomic levels.

Energy of excitation of atom can be removed, also due to the emission of Auger electrons. These electrons are emitted instead of X-ray quantum and possess energy equal to the energy of X-ray quantum minus bound energy of electron at the corresponding level

$$E_{ea} = E_x - E_{bi} \tag{40}$$

where E_{ea} – energy of Auger electron.

In Table 11 are given both the most intensive X-ray and gamma-lines their relative escape on 100 formed vacancies [49]. We have made calculation of X-ray quanta on 100 captured neutrons (the data are presented in the table also).

Table 12.

X-Ray Energy (keV)	Shell Comments	Intensity (1/100 Vacancies)	Intensity [1/100 neutrons]
181.931	Gamma		139(6)
79.510	Gamma		77.3(19)
42,996	Kα1	47,5	28,5
42,309	Kα2	26,6	15,96
41,864	Kα3	0,00824	0,004944
48,695	Kβ1	9,3	5,58
49,959	Kβ2	3,11	1,866
48,551	Kβ3	4,81	2,886
50,099	Kβ4	0,038	0,0228
49,038	Kβ5	0,146	0,0876
50,219	KO2,3	0,45	0,27
6,058	Lα1	6,3	3,78
6,026	Lα2	0,7	0,42
6,713	Lβ1	3,9	2,34
7,102	Lβ2,15	1,32	0,792
6,832	Lβ3	0,126	0,0756
6,687	Lβ4	0,077	0,0462
7,243	Lβ5	0,0108	0,00648
6,867	Lβ6	0,063	0,0378
7,79	Lγ1	0,67	0,402
8,087	Lγ2	0,024	0,0144
8,105	Lγ3	0,034	0,0204
7,93	Lγ6	0,0055	0,0033
6,049	Lη	0,092	0,0552
5,362	L1	0,266	0,1596

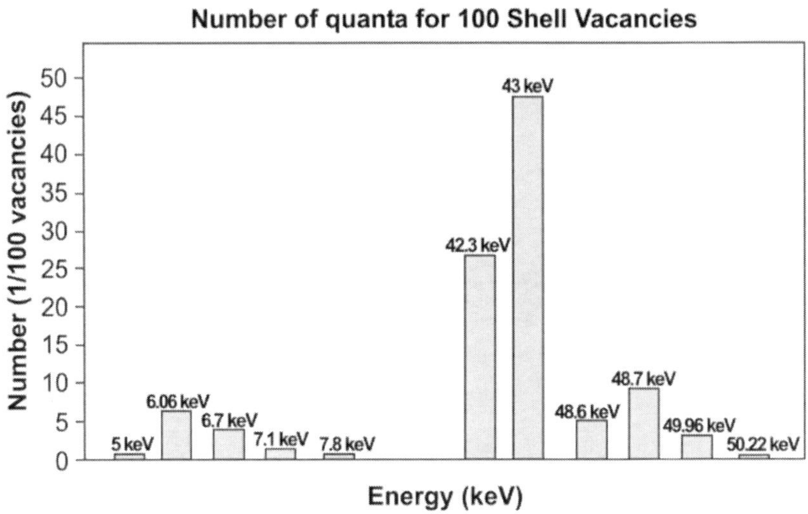

Figure 24. Amount of X-ray photons depending on their energy, which formed, on 100 vacancies in electron subshells of Gadolinium.

The possibility output of X-ray quanta on 100 formed vacancies is represented in the Figure 24. From figure one can see, that the most intensive are K-lines 42.3 and 43 keV. The most interesting for us are low-energy L-lines however their output is strongly suppressed by Auger electrons irradiation.

During the passage of X-ray quanta through a material of the converter there absorption occurs. The degree of absorption is described by the mass factor μ/ρ of attenuation of X-rays.

Attenuation of narrow beam of monoenergetic photons with an incident intensity I_o, penetrating a layer of material with mass thickness x and density ρ, emerges from material with intensity I occur by the exponential attenuation law

$$I/I_o = \exp[-(\mu/\rho)x] \; . \tag{41}$$

Equation (41) can be rewritten as

$$\mu/\rho = x^{-1}\ln(I_o/I) \tag{42}$$

μ/ρ can be obtained from measured values of I_o, I and x.

Note that the mass thickness is defined as the mass per unit area, and is obtained by multiplying the thickness t by the density ρ, i.e., $x = \rho t$.

Present tabulations of μ/ρ rely based on theoretical values for the total cross section per atom, σ_{tot}, which is related to μ/ρ according to

$$\mu/\rho = \sigma_{\mathbf{tot}}/uA\ . \tag{43}$$

In (eq. 43), $u = (1.6605402 \times 10^{-24}$ g) [51] is the atomic mass unit (1/12 of the mass of an atom of the nuclide ^{12}C), A is the relative atomic mass of the target element, and σ_{tot} is the total cross section for an interaction by the photon, frequently given in units of b/atom (barns/atom), where $b = 10^{-24}$ cm^2.

The attenuation coefficient of photon interaction cross sections and related quantities are functions of the photon energy. The total cross section can written as the sum contributions of various principal photon interactions modes,

$$\sigma_{\mathbf{tot}} = \sigma_{\mathbf{pe}} + \sigma_{\mathbf{coh}} + \sigma_{\mathbf{incoh}} + \sigma_{\mathbf{pair}} + \sigma_{\mathbf{trip}} + \sigma_{\mathbf{ph.n.}} \tag{44}$$

where σ_{pe} is the atomic photo effect cross section, σ_{coh} and σ_{incoh} are the coherent (Rayleigh) and the incoherent (Compton) scattering cross sections, respectively, σ_{pair} and σ_{trip} are the cross sections for electron-positron production in the fields of the nucleus (pair) and with addition radiation of the atomic electrons (triple), respectively, and $\sigma_{ph.n.}$ is the photonuclear cross section.

Accordingly μ/ρ can represented as following

$$\mu/\rho = (\sigma_{\mathbf{pe}} + \sigma_{\mathbf{coh}} + \sigma_{\mathbf{incoh}} + \sigma_{\mathbf{pair}} + \sigma_{\mathbf{trip}})/uA\ . \tag{45}$$

Values of the mass attenuation coefficient μ/ρ for the mixtures and compounds (assumed non homogeneous) were obtained according to simple additive:

$$\mu/\rho = \sum_i w_i (\mu/\rho)_i \qquad (46)$$

where w_i is the fraction by weight of the ith atomic constituent.

The methods used to calculate the mass energy-absorption coefficient μ_{en}/ρ are described perhaps more clearly to the use of an intermediate quantity, the mass energy-transfer coefficient, μ_{tr}/ρ.

Mass factors of absorption of X-ray quanta for various materials are represented in the tabulated data [51]. On the Figure 25 the dependence of μ/ρ on energy of quanta is depicted. In figure peaks and edges of absorption of electronic subshells K, L, M are well visible.

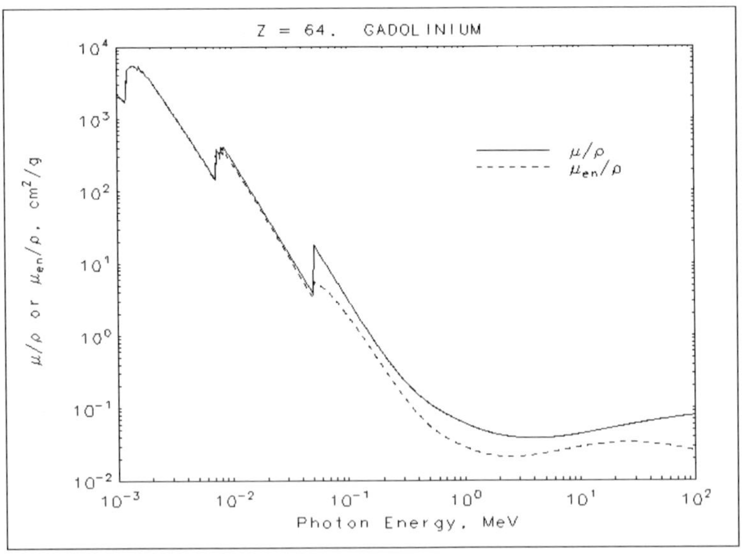

Figure 25. Dependence of mass factor of absorption of quanta on their energy for gadolinium [52].

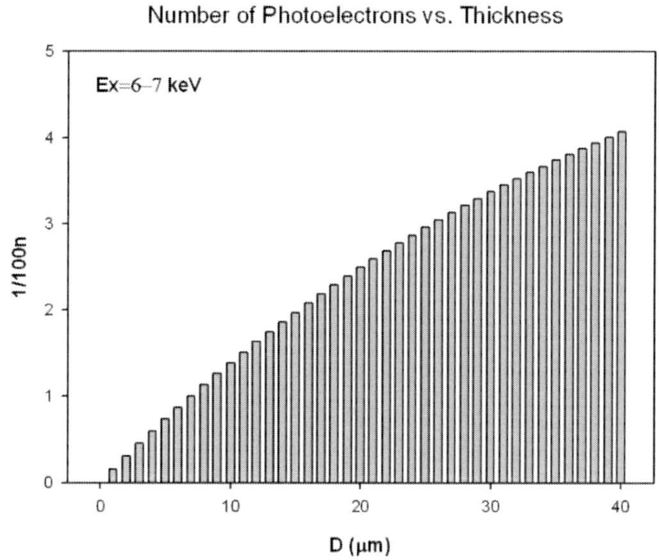

Figure 26. Dependence of number of formed electrons on thickness of the converter. For quanta with the energy range 6 up to 7 keV for 100 falling neutrons.

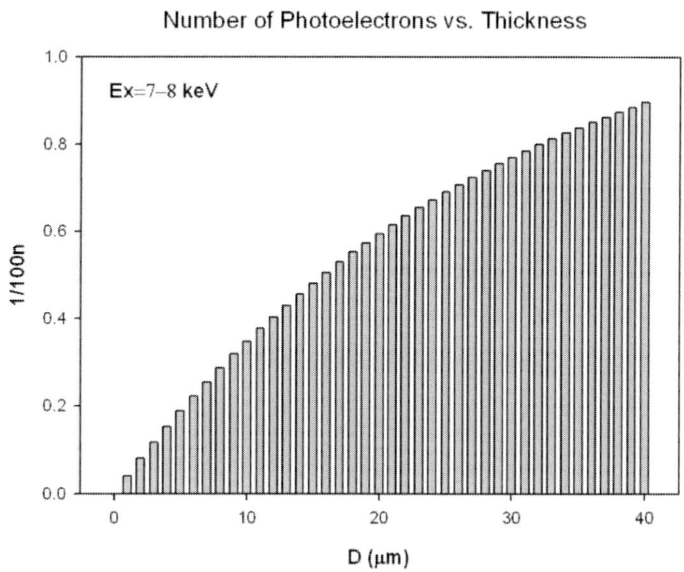

Figure 27. Dependence of number of formed electrons on thickness of the converter. For quanta with the energy range from 7 up to 8 keV for 100 falling neutrons.

Absorption of quanta strongly depends on their energy. Low-energy quanta are absorbed most intensively, however formed secondary electrons are also the low-energy ones and have rather small run in a material of the converter.

In the below figures dependences of formation of secondary electrons on depth of penetration of quanta's are given. Amounts of formed electrons are given without taking into account their absorption in a material of the converter (an integral one).

In Figure 28 the integrated characteristic of formation of secondary electrons is presented. The low-energy quanta's form mainly photoelectrons, and gamma-quantum with the energy 181.9 keV forms Compton electron. The low energy quanta's with the energy range from 5 and up to 9 keV form low-energy electrons, the maximal free length path of which in gadolinium are accordingly equal to 0.4 and 0.6 microns. At calculation of the escaping electrons only the quanta's formed in a layer, which is a little thicker, than a maximal free length path of electrons for the given energy, are taken into account. In the table the new data, of electrons which are formed as a result of X-ray and low-energy quanta absorption shown. These electrons are designated in the comment as X.

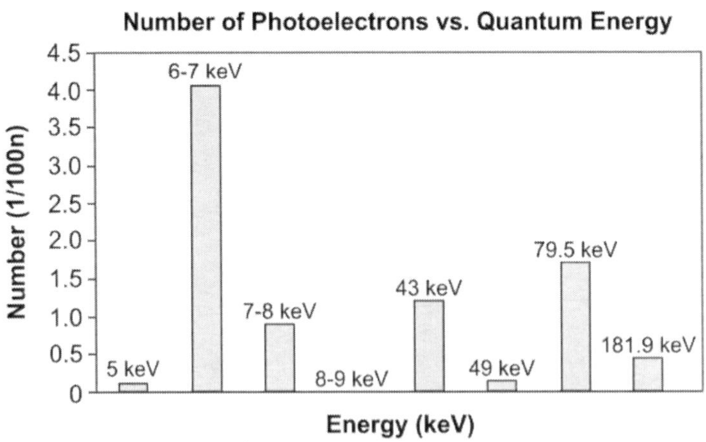

Figure 28. Dependence of the number of secondary electrons depending on energy of quanta for 100 falling neutrons.

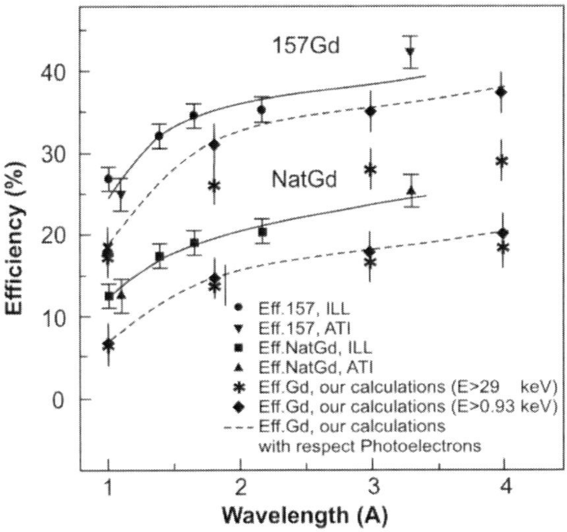

Figure 29. Comparisons of our theoretical calculations with the experimental data give in the paper [14]. Calculations are carried out for two boundary energies taken into account electrons; more than 29 keV and more than 0.93 keV and in view of the contribution of secondary electrons formed by X-ray and low-energy gamma-quanta (a dashed line).

The contribution of photoelectrons to a total sum of electrons for each value of energy does not exceed 2%. And their total amount does not exceed 0.97 electrons on 100 incident neutrons that is 1% less.

Calculations of the contribution of photoelectrons in general efficiency of gadolinium converters are carried out. Results of calculations give in Figure 29.

From the figure one can see that in the case of the account of the contribution of electrons formed by X-ray quanta, the efficiency is increased a little, but their contribution is insignificant. The increase occurs for no more than by 1%. It is rather possible, that X-ray quanta can essentially effect on the total efficiency at the account of their absorption in a material of detectors. So gas detectors with Argon filling can increase efficiency essentially. If the conversion of quanta will occur in a working body of the detector, therefore the full gathering of formed secondary electrons will take place. They can render even greater influences in the case of use of semi-conductor detectors.

Table 13.

Energy of electrons (keV)	Electron escape 1/100 n (error)	Electron run in gadolinium (μm)	Energy of initial gamma-quantum	Remarks, level
0.97	>200	0,04		M-Auger
4.84	97(33)	0.3		L-Auger
5 - 8	0,2	0,4 - 0,6		X
29.3	35,58	4,7	79.51	K
29,3	0,18	4,7		X
34.9	7.9(4)	6,29		K- Auger
34,9	0,18	6,29		X
71.7	5,57	20,7	79.51	L
78	1,2	23,78	79.51	M
78	0,03	23,78		X
131.7	6,96	55,70	181.93	K
131,7	0,38	55,70		X
174.1	0,99	86,27	181.93	L
180.4	0,21	91,23	181.93	M
205.4	0,14	111,47	255.66	K
227.3	0,16	130,27	277.54	K
729.9	0,03	649,38	780.14	K
893.85	0,06	830,05	944.09	K
911.8	0,04	849,83	960	K
926.8	0,03	866,35	975.4	K

2.2.5. Modeling of Converters Representing Sandwiches, From Supporting Films and Converters

Recently more often started to apply widely variety solid-state converters of thermal neutrons of complicated forms, such as bilateral converters made from thin gadolinium with supporting film made from kapton, neutron converters on the base of GEM-structures and others. Manufacturing techniques of this converters consist in putting of thin gadolinium converters (0.5-5 microns) on the surfaces of film made from kapton with the thickness of 20-100 microns. Such converters allow manufacturing detectors of the big areas. Presence of supporting films results of additional reduction of the secondary escaping electron, at the same time the use of bilateral converters allows to increase efficiency of detection of neutrons, at use of two detectors from both sides of the converter. Modeling of efficiency of registration of thermal neutrons by these kinds of converters is an interesting problem and during its development one can expect occurrence of new converters.

Modeling of the efficiency of registration of thermal neutrons by the converters, which represent kapton film, serving for support of thin gadolinium layers is carried out. Layers can be settled down on one, or on both sides of supporting film. Calculations are made for natural gadolinium and its 157 isotope, for four fixed energies of neutrons. In this calculations electrons with energy more than 29 keV were considered only.

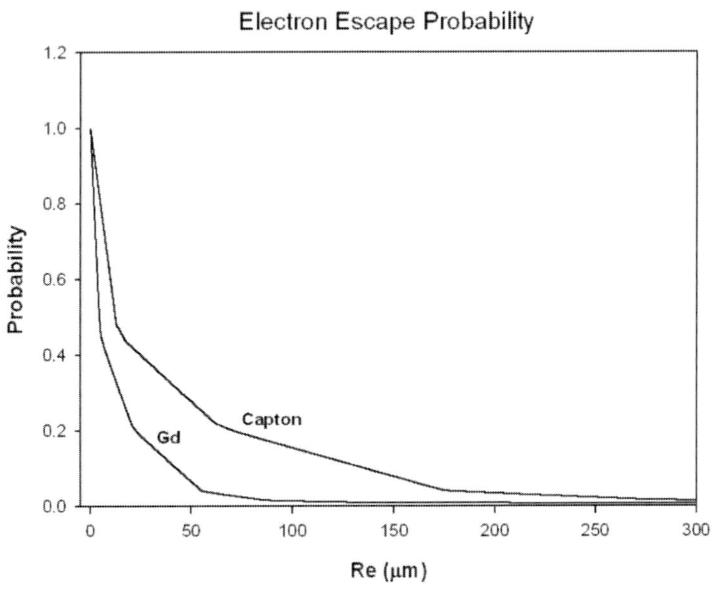

Figure 30. Probability of electron output, formed in the reaction of radiating capture of thermal neutrons by gadolinium nuclei, from kapton and gadolinium.

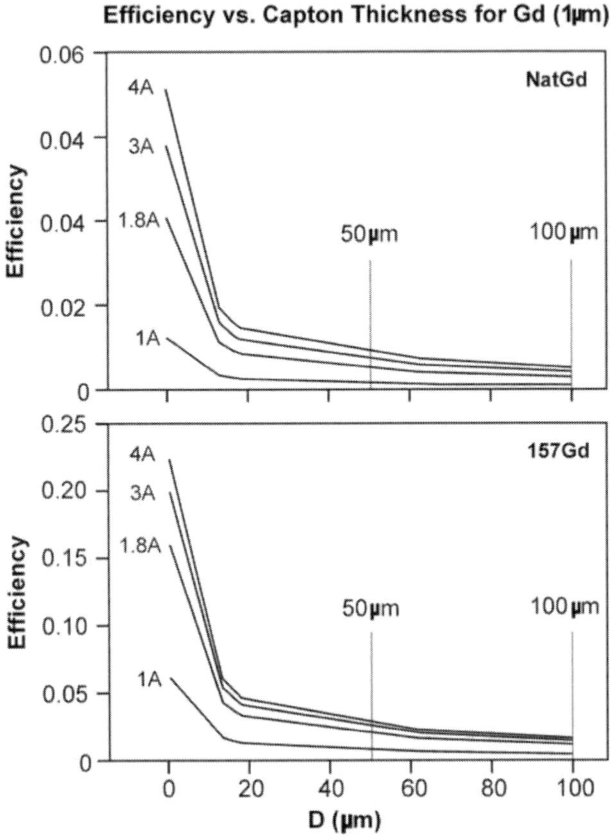

Figure 31. Efficiency of complex converters made from thin gadolinium (1 micron) put on a substrate from kapton, depending on thickness of a substrate. Various curves characterize wavelengths of neutrons.

Attenuation of electron flux arising in the reaction of radiating capture of neutrons by gadolinium nucleus occurs both in the substance of the converter and in kapton film. In Figure 30 curves of probabilities of electron flux output for gadolinium and kapton are given. These curves characterize absorption of electrons in these substances.

With contributions of various energies of electrons which are well visible, it is conditionally possible to separate it into 4 groups. In first group the least energetic and having small run in substance in the fourth most energetic electrons, Figure 4.

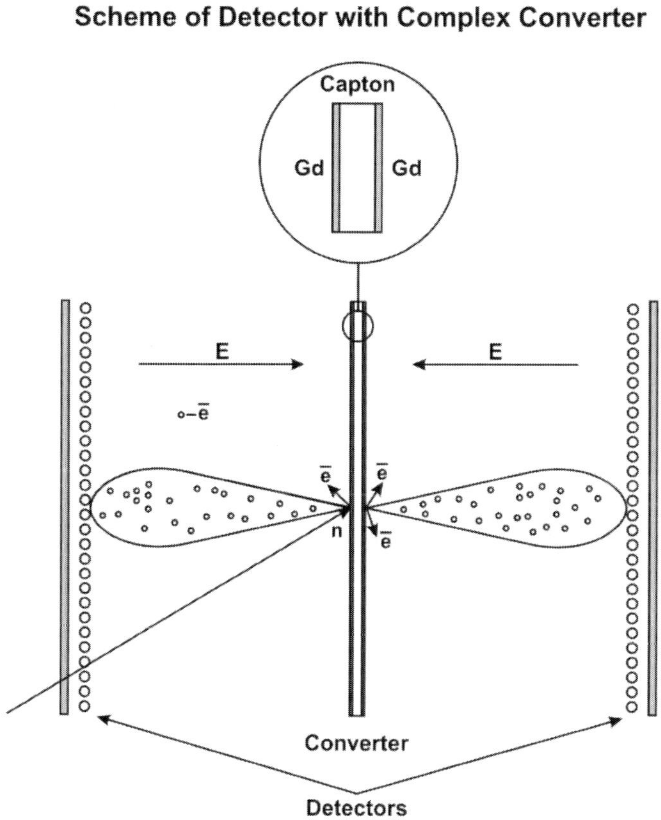

Figure 32. The scheme of the detector used in article [52]. The converter consists of a supporting film executed of kapton film from which two sides of its two gadolinium converter were placed.

Calculations of efficiency of converters with the thickness of 1 micron made from natural gadolinium and its 157 isotope, for four fixed neutron wavelengths are carried out. Calculations were performed for the case, when only one thin converters were directly settled on the films made from kapton of various thicknesses, Figure 31. This kind of geometry is characteristic for a case of drawing of thin converting films (0.5-3 microns) on supporting substrates made from kapton. Thickness of supporting films as usual is chosen within the limits of 20-100 microns.

Converters like this was developed and used in DETNI project in Hahn-Meitner-Institut (HMI). The detector and converter are described in the paper [52].

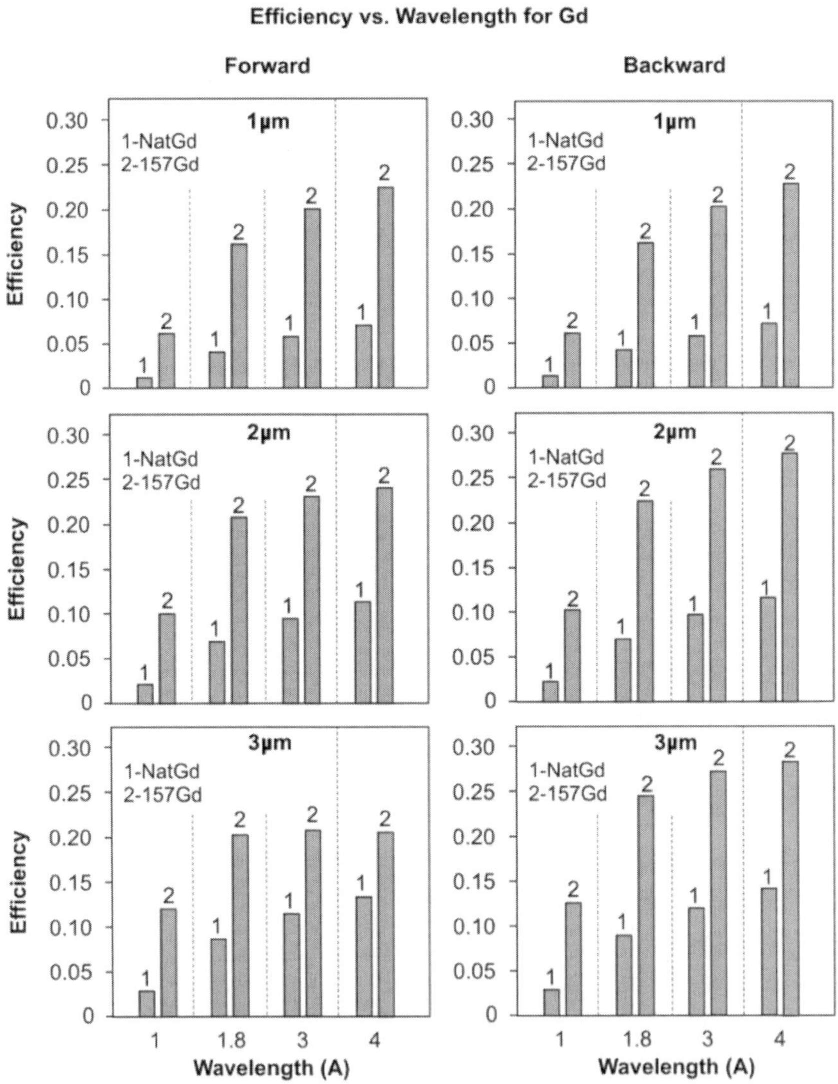

Figure 33. Efficiency of registration of neutrons with wavelengths 1, 1.8, 3 and 4 Å for ^{nat}Gd and ^{157}Gd for converters with the thickness 1, 2 and 3 microns without supporting films.

The detector's size is 285 x 285 мм^2 shown on the Figure 32. Each segment consists of centrally located converter consisting of ^{157}Gd/CsI foil. Converters are located from both sides of supporting foil. Thicknesses of converters lay in the range from 0.5 - up to 1.5 microns of ^{157}Gd. CsI is used for thermalization of electron energy and as an emitter of secondary electrons, has thickness less than 1 micron. Necessity of the use of secondary emitter CsI caused by the reason that there plane-parallel chambers of low pressure are used as detector. Electrons escaping from the converter have enough big range in low pressure gases, which can result in appreciable deterioration of the spatial resolution.

Calculations were carried out for converters made from natural gadolinium and its 157 isotope, for 4 fixed neutron wavelengths. Thickness of converters varied from 1 up to 3 microns. Calculations was made for only electrons with energy more than 29 KeV. At calculations the secondary emitters of electrons executed on the basis of crystals CsI were not considered.

In Figure 33 effects of calculations of efficiency of gadolinium foil in simple geometry are given. Calculations were made for one foil without supporting films. Efficiency for converters from natural gadolinium grows with enlargement of a thickness. At use converters from 157 isotopes of gadolinium efficiency for electrons departs in a back hemisphere not significant growth. For electrons departing for a forward hemisphere even efficiency decrease is observed.

For this complicated converter, at an arrangement of the converters from both sides of supporting film, the part of neutrons will be converted in the first converter and a part in the second. From the first converter electrons emitting to the back hemisphere will be registered, by the forward chamber. From the second converter electrons emitting to the forward hemisphere will be registered by last chamber. These processes will be the primary. At the same time a part of high energy electrons can penetrate through the supporting film and will be registered by the next chamber. This process can increase total efficiency of registration a little.

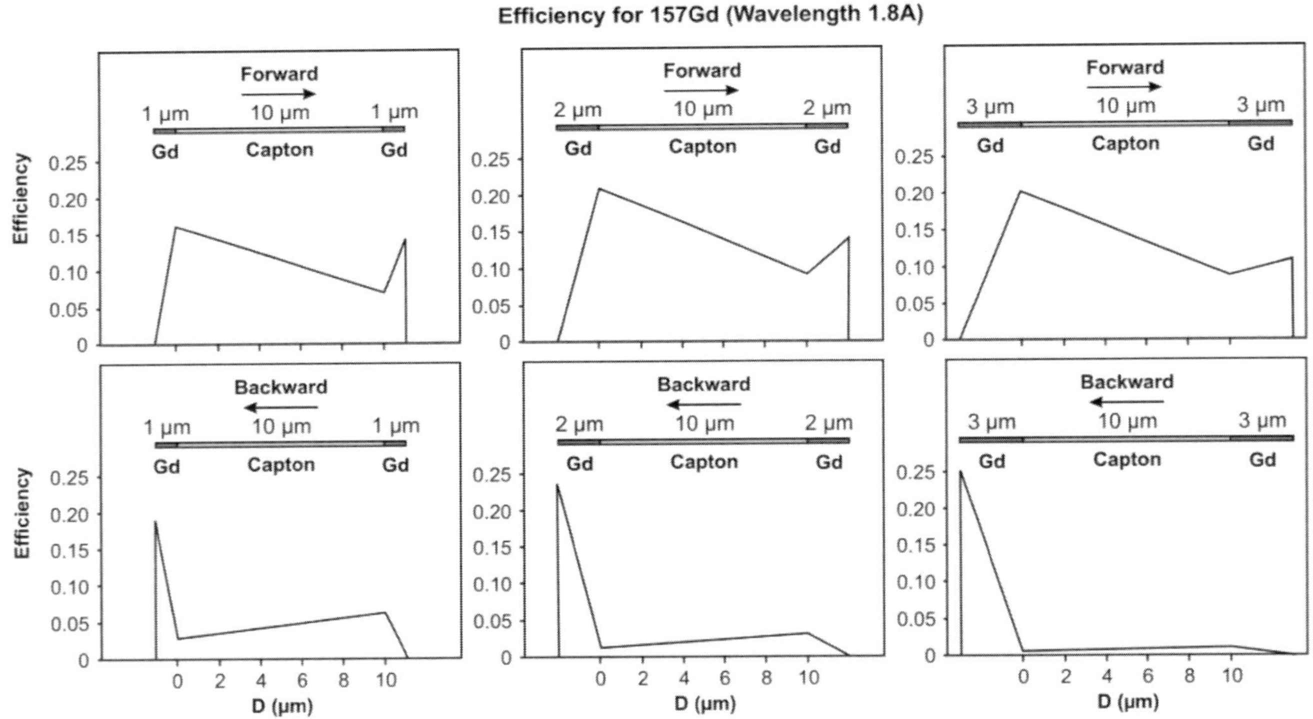

Figure 34. Efficiency of the complex converter for a case of electron output to the forward and back hemispheres is given. Thickness of the converter from ^{157}Gd is 1, 2 and 3 microns, thickness of kapton is 10 microns. Wavelength of neutrons is 1.8 Å.

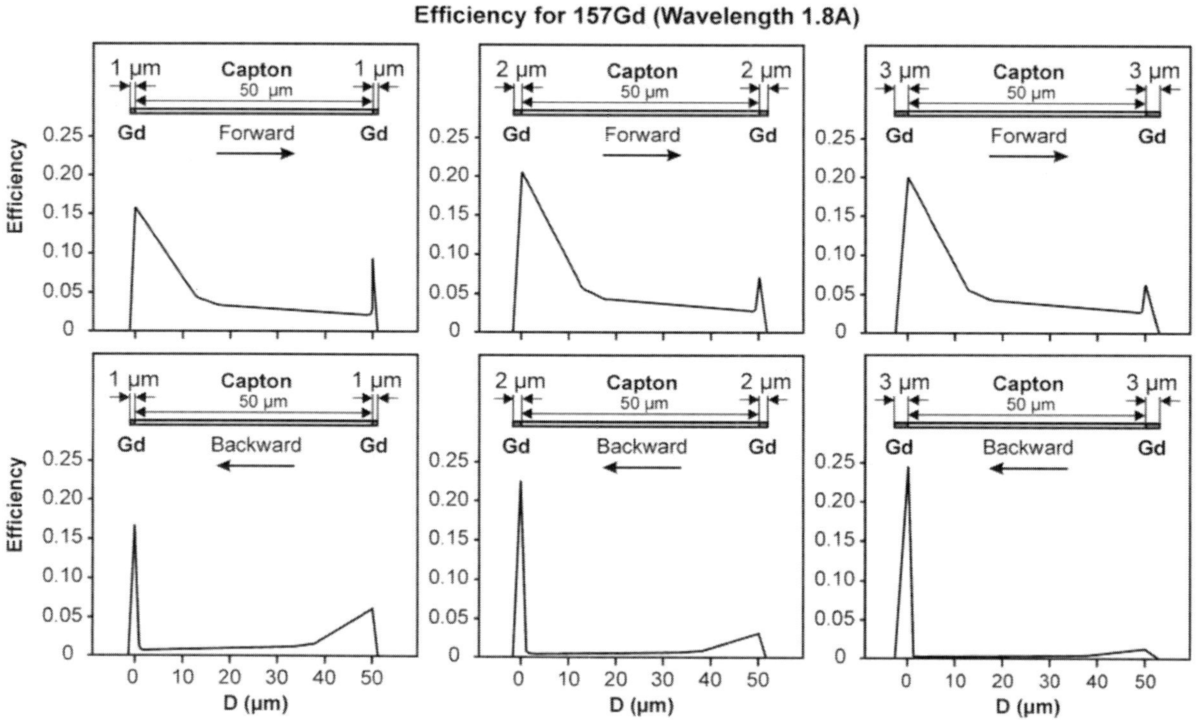

Figure 35. Efficiency of the complex converter for a case of electron output to the forward and back hemispheres is given. Thickness of the converter from ^{157}Gd is 1, 2 and 3 microns, thickness of kapton is 50 microns. Wavelength of neutrons is 1.8 Å.

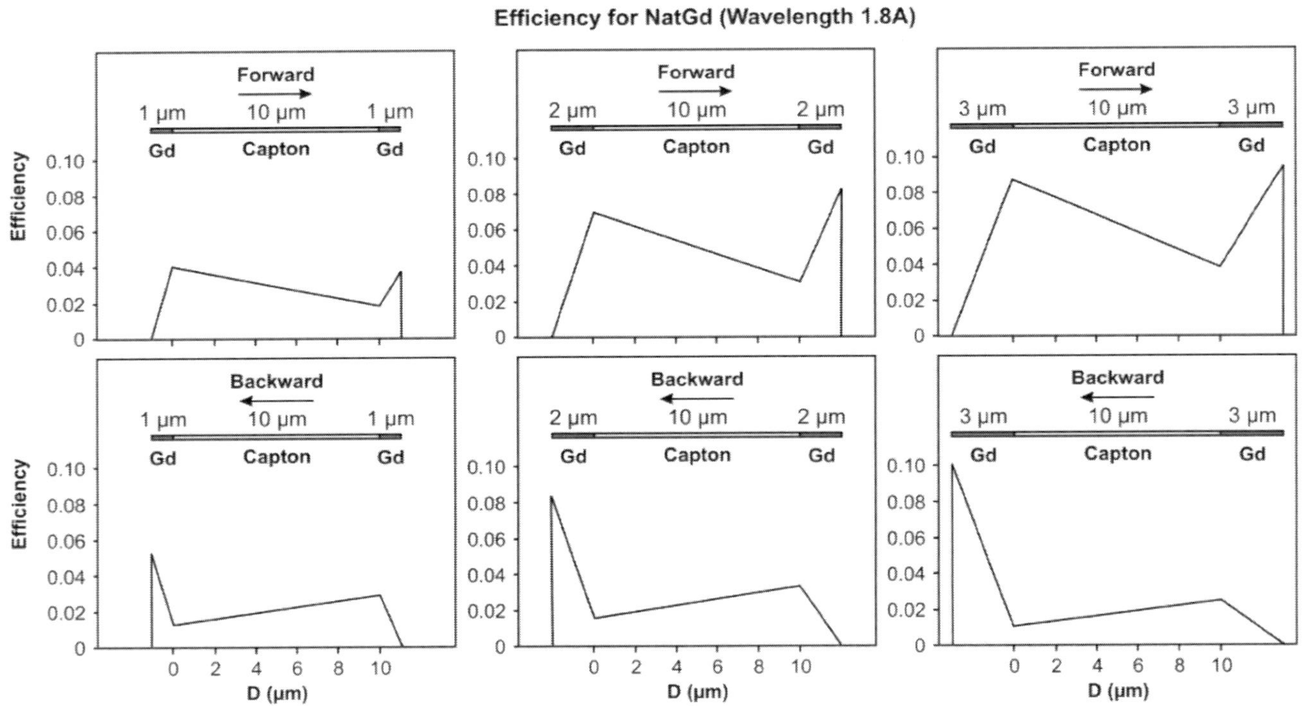

Figure 36. Efficiency of the complex converter for a case of electron output to the forward and back hemispheres is given. Thickness of the converter from ^{nat}Gd is 1, 2 and 3 microns, thickness kapton is 10 microns. Neutron wavelength is 1.8 Å.

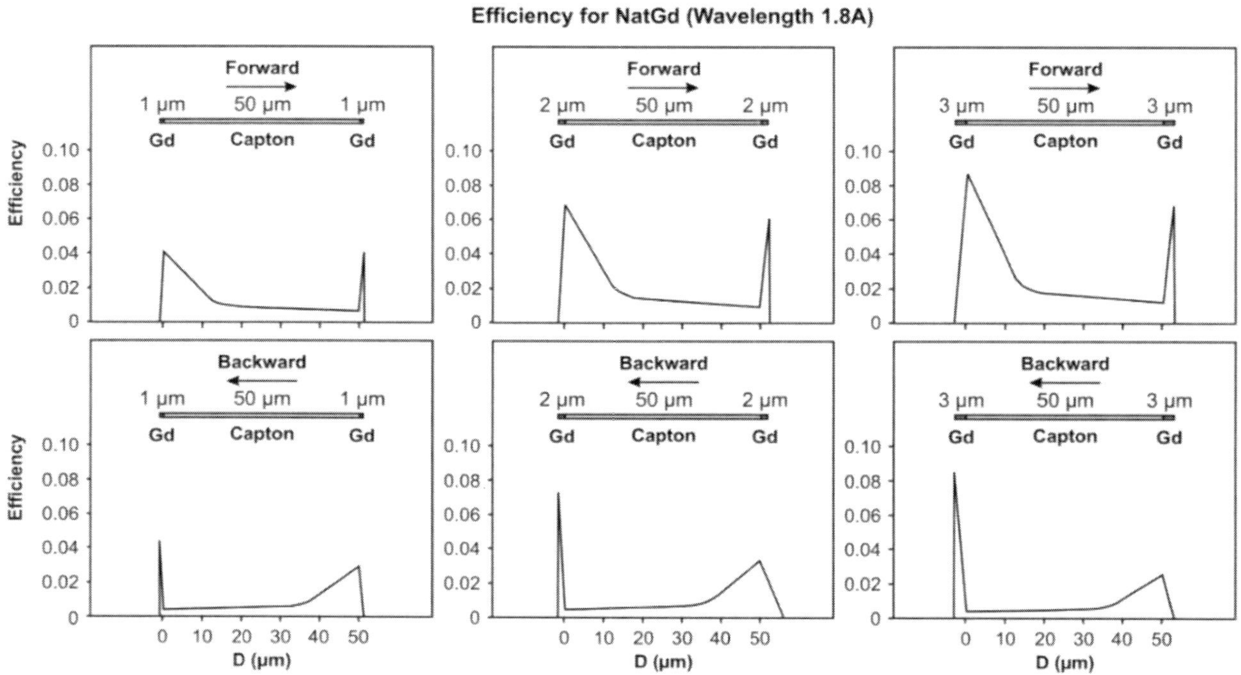

Figure 37. Efficiency of the complex converter for a case of electron output to the forward and back hemispheres is given. Thickness of the converter from ^{nat}Gd is 1, 2 and 3 microns, thickness kapton is 50 microns. Neutron wavelength is 1.8 Å.

Table 14.

Kapton Thickness, microns	Geometry, microns ^{157}Gd	Contribution of converters in the efficiency (forward electron emission) last detector			Total of both converters
		1 layer	Escape from the 1 layer in view of attenuation	2 layer	
10	1 + 1	0.160	0.070	0.074	0.144
	2 +2	0.208	0.091	0.045	0.136
	3 + 3	0.202	0.089	0.020	0.109
20	1 + 1	0.160	0.032	0.074	0.106
	2 +2	0.208	0.042	0.045	0.087
	3 + 3	0.202	0.041	0.020	0.061
30	1 + 1	0.160	0.028	0.074	0.102
	2 +2	0.208	0.037	0.045	0.082
	3 + 3	0.202	0.036	0.020	0.056
40	1 + 1	0.160	0.025	0.074	0.099
	2 +2	0.208	0.032	0.045	0.077
	3 + 3	0.202	0.031	0.020	0.051
50	1 + 1	0.160	0.021	0.074	0.095
	2 +2	0.208	0.027	0.045	0.072
	3 + 3	0.202	0.026	0.020	0.046

Table 15.

Kapton Thickness, microns	Geometry, microns ^{157}Gd	Contribution of converters in the efficiency (back electron emission) forward detector			Total of both converters
		1 layer	2 layer	Escape from the 2 layer in view of attenuation	Total of both converters
10	1 + 1	0.161	0.062	0.027	0.188
	2 +2	0.223	0.031	0.014	0.237
	3 + 3	0.246	0.011	0.005	0.251
20	1 + 1	0.161	0.062	0.012	0.173
	2 +2	0.223	0.031	0.006	0.229
	3 + 3	0.246	0.011	0.002	0.248
30	1 + 1	0.161	0.062	0.011	0.172
	2 +2	0.223	0.031	0.005	0.228
	3 + 3	0.246	0.011	0.002	0.248
40	1 + 1	0.161	0.062	0.010	0.171
	2 +2	0.223	0.031	0.005	0.228
	3 + 3	0.246	0.011	0.002	0.248
50	1 + 1	0.161	0.062	0.008	0.169
	2 +2	0.223	0.031	0.004	0.227
	3 + 3	0.246	0.011	0.001	0.247

Table 16.

Kapton Thickness, microns	Geometry, microns ^{nat}Gd	Contribution of converters in the efficiency (forward electron emission) last detector			Total of both converters
		1 layer	Escape from the 1 layer in view of attenuation	2 layer	Total of both converters
10	1 + 1	0.041	0.018	0.035	0.053
	2 +2	0.069	0.030	0.052	0.082
	3 + 3	0.087	0.038	0.056	0.094
20	1 + 1	0.041	0.008	0.035	0.043
	2 +2	0.069	0.014	0.052	0.066
	3 + 3	0.087	0.017	0.056	0.073
30	1 + 1	0.041	0.007	0.035	0.042
	2 +2	0.069	0.012	0.052	0.064
	3 + 3	0.087	0.015	0.056	0.071
40	1 + 1	0.041	0.006	0.035	0.041
	2 +2	0.069	0.011	0.052	0.063
	3 + 3	0.087	0.013	0.056	0.069
50	1 + 1	0.041	0.005	0.035	0.04
	2 +2	0.069	0.009	0.052	0.061
	3 + 3	0.087	0.011	0.056	0.067

Calculation of efficiency for thin gadolinium converters is made in view of their attenuation in kapton. Thus thickness of the kapton film (from 10 up to 50 microns) and thickness of gadolinium (from 1 up to 3 microns) were varied. From the Figure 34 and 35 one can see, that the optimal thickness for an electron emission to the back hemisphere is gadolinium 157 thickness of 2 microns. Up to the thickness of 3 microns growing of efficiency is observed, however this converter is a lobby and the further increase of its thickness will result in shielding the second converter. For the second converter (electron emission to the forward hemisphere) an optimum thickness is 2 microns as the further increase of thickness does not result in increase of efficiency. The optimal thickness for a kapton film is the thickness of 10 microns. With the growth of thickness the secondary escape of electron falls. Especially it has an effect at registration electrons emitted to the forward hemisphere.

Table 17.

Kapton Thickness, microns	Geometry, microns ^{nat}Gd	Contribution of converters in the efficiency (back electron emission) forward detector			Total of both converters
		1 layer	Escape from the 2 layer in view of attenuation	2 layer	Total of both converters
10	1 + 1	0.040	0.013	0.029	0.053
	2 +2	0.069	0.014	0.033	0.083
	3 + 3	0.089	0.011	0.025	0.1
20	1 + 1	0.040	0.006	0.029	0.046
	2 +2	0.069	0.007	0.033	0.076
	3 + 3	0.089	0.005	0.025	0.094
30	1 + 1	0.040	0.005	0.029	0.045
	2 +2	0.069	0.006	0.033	0.075
	3 + 3	0.089	0.004	0.025	0.093
40	1 + 1	0.040	0.004	0.029	0.044
	2 +2	0.069	0.005	0.033	0.074
	3 + 3	0.089	0.004	0.025	0.093
50	1 + 1	0.040	0.004	0.029	0.044
	2 +2	0.069	0.004	0.033	0.073
	3 + 3	0.089	0.003	0.025	0.092

Figures show that at use of 157 isotope of gadolinium the main absorption of neutrons occurs on the forward converter. The second converter is substantially shielded by the first converter.

Similar calculations conducted also for converters made from natural gadolinium; results of these calculations presented in Figures 36 and 37.

In Table 13 contributions of each of 2 converters in the efficiency of the last (2 detector) detector, an electron emission in the forward hemisphere are shown. Converters made from 157 gadolinium isotope. Neutron wave length is 1.8 Å.

In Table 14 contributions of each of 2 converters in the efficiency of the forward detector (1 detector), an electron emission to the back hemisphere are presented. Converters made from 157 gadolinium isotope. Neutron wave length is 1.8 Å.

In Table 15 contributions of each of 2 converters in the efficiency of the last detector, an electron emission to the forward hemisphere are presented. Converters made from natural gadolinium. Neutron wave length is 1.8 Å.

In Table 16 contributions of each of 2 converters in the efficiency of the forward detector, an electron emission to the back hemisphere are given. Converters made from natural gadolinium. Neutron wave length is 1.8 Å.

2.2.6. Modeling of Converters Executed from a Set Thin Drilling Converters

The new solid-state converter for thermal neutrons was offered. The converter consist of a set of thin gadolinium foils located one over other in a same gas volume. Foils there will be drilled with the fine step (2 mm) with diameter of apertures 1 mm. These foils will have an optical transparency of 40%; correspondingly, gadolinium will fill 60% of a surface. Secondary electrons will emit for all sides, and in an electric field they will be entice down in holes and further drift in the direction of the detector. As a detector there a various gas detectors, such as the multi-wire proportional chamber, multi-step avalanche and multi-strip detectors, and so on, can be used. The schematic sketch of this kind of detector is shown on the Figure 38. On the Figure 39 the cross-section of the detector is shown.

Technologically this kind of converter can make from foils with the thickness more than 5 microns. However, more thin foils are also having interest for calculations. The distance between the foils can be chosen in a range from 0.2 up to 2 mm.

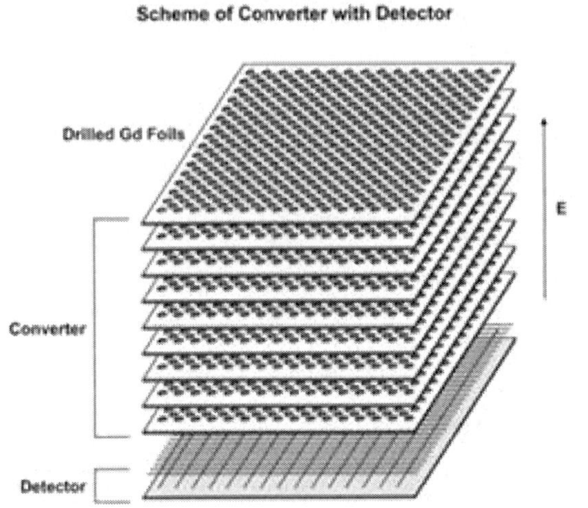

Figure 38. A schematic sketch of the detector consisting of the complex converter made from drilled by fine step foils located one over other in the same gas volume.

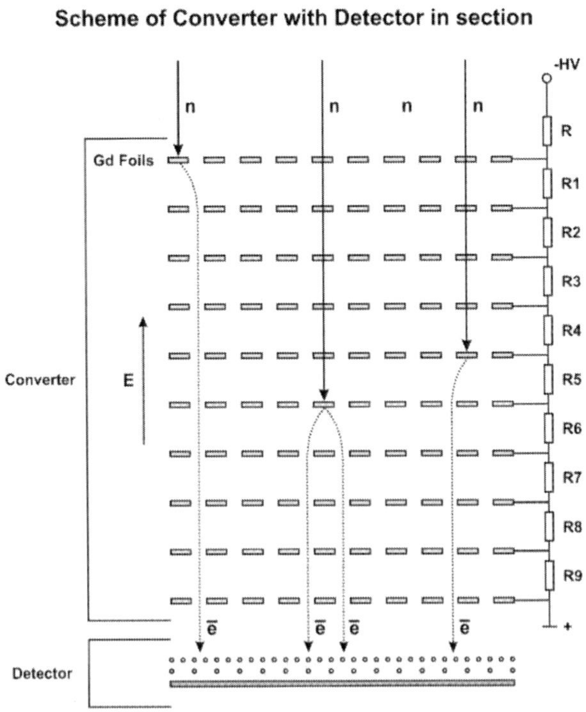

Figure 39. Scheme of the complex detector.

For avalanche detectors working at normal atmospheric pressure the distance of 0,2 mm will be enough for secondary electron formation. Secondary electrons, having small energy, will drift on the direction of electric field to the detector. Selecting an intensity of the electric field it is possible to achieve an effective transfer of electrons to the drift gap and further to the detector.

This kind detector will have not so good spatial resolution, which will be modulated with the step of apertures (in this case 2 mm). But for the series of tasks this resolution is sufficient. In order to improve the spatial resolution one should prepare foils with finer step, correspondingly, with the smaller diameter of apertures.

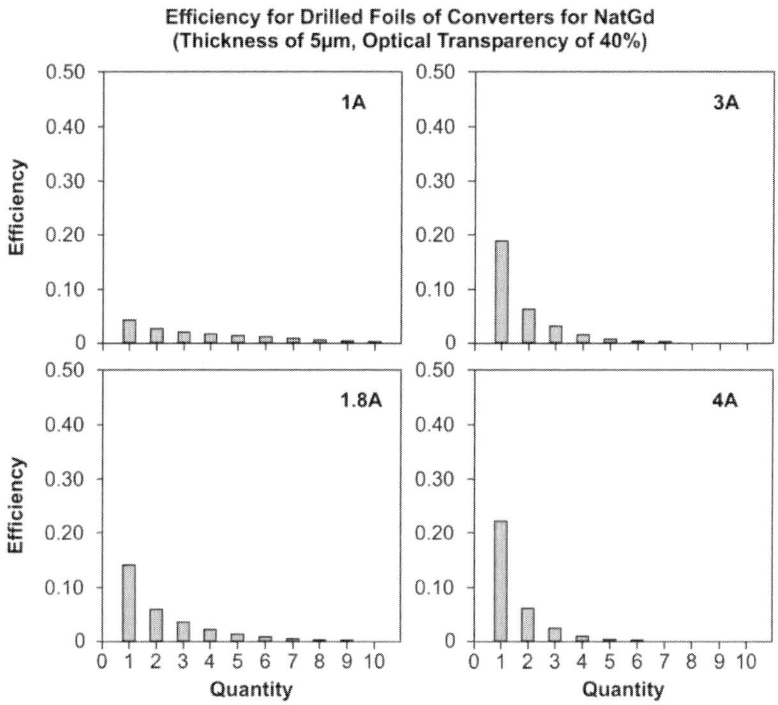

Figure 40. Efficiency of every one of the subsequent drilled converter from 5 microns thickness natural gadolinium with an optical transparency of 40%.

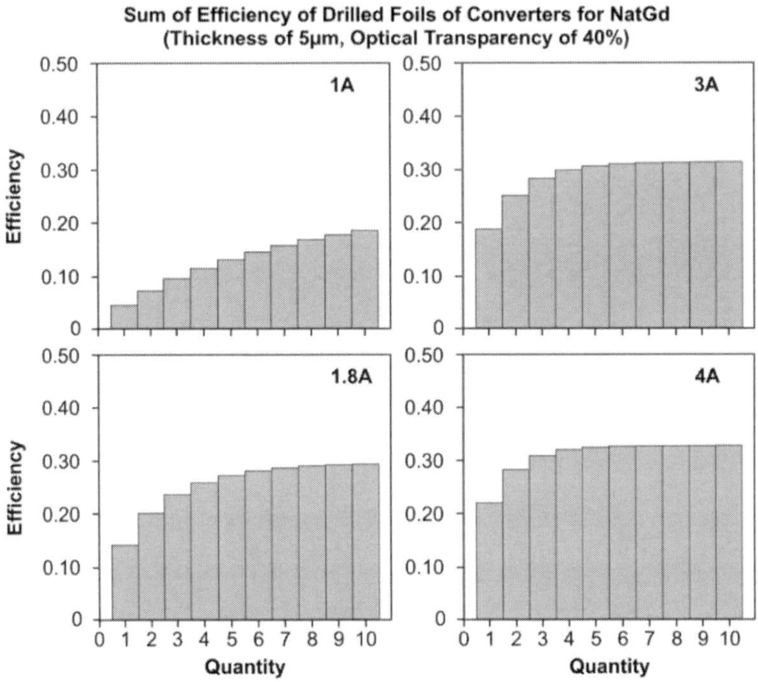

Figure 41. Summary efficiency of the converter consists from 10 drilled foils of natural gadolinium with a thickness of 5 micron.

Calculations of this kind collimator are carried out. In calculations only conversion and Auger electrons with the energy higher than 29 keV isotropic emitting to the all sides were considered. Efficiency for each foil and their sum were separately calculated. In the figures 40 and 42 results of calculations for the converters made from ^{nat}Gd with the thicknesses of 5 and 10 microns correspondingly, presented. Efficiency is calculated for each foil separately. Absorption of neutrons can occur not only at forwards foils, but at subsequent foil too. The greatest efficiency can be received for foils with the thickness of 1 micron. At use the foil with the thickness more than 5 microns the efficiency decreases due to a loss of secondary escaping electron in the body of converters. In Figures 41, 43 total efficiency such converter is shown. For calculations is in interest only converters from natural gadolinium as it can increment their total efficiency.

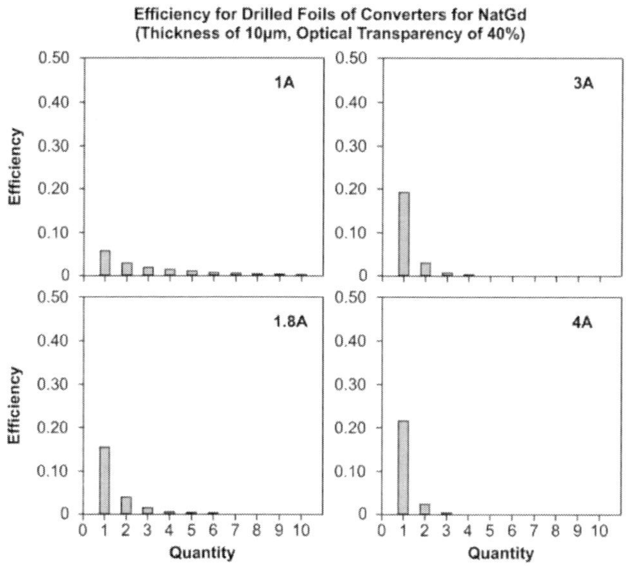

Figure 42. Efficiency of every one of the subsequent drilled converter from10 microns thickness natural gadolinium with an optical transparency of 40%.

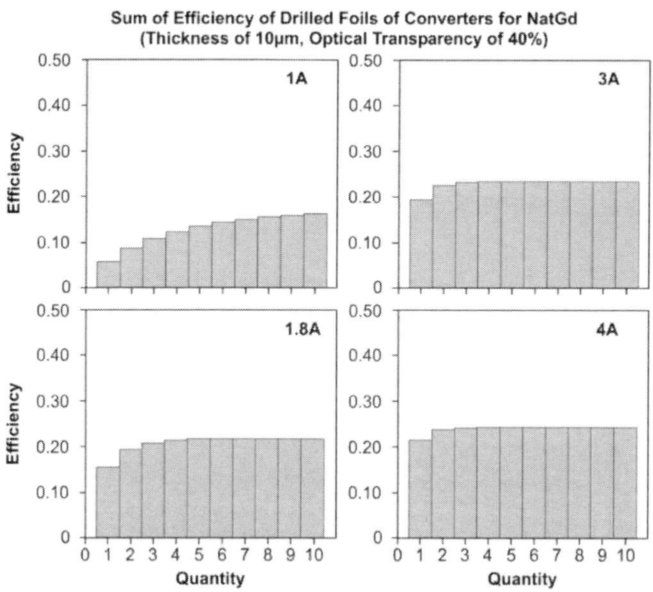

Figure 43. Summary efficiency of the converter consists from 10 drilled foils of natural gadolinium with a thickness of 10 micron.

Chapter 3

Position Sensitive Detectors of Thermal Neutrons with Gadolinum Converters

Neutron imaging was at first successfully used in 1935, shortly after the discovery of the neutron by James Chadwick. At present, neutron imaging methods are widely used in a number of institutions. First of all in the Paul Csherrer Institut (PSI) [53], as well as in such institutes as French Atomic Commission (CEA) [54], Hahn-Meitner-Institut (HMI) [55], ISIS in the UK [56], in Norway, Kjeller Institute for Energy Technology [57], at IBR-2 in Dubna, Russia [58], in the Czech Republic Rez reactor [59]. There are of course plans to build a neutron imaging facility at the ESS and the ILL in Grenoble is considering building a new neutron imaging facility.

In installations for neutron imaging gadolinium foils are used as neutron radiation converters. In the beginning, X-ray films were widely used as detectors, after the thin-layer scintillators with photos and movie cameras began to be used. With the development of technology appeared installations based on the SSD and a number of electronic detectors was offered.

Technology for neutron imaging is constantly evolve, new types of detectors with better characteristics are develop.

The idea of using thin foils to convert neutrons radiation into charged particles and count the conversion products in a solid-state detector was originally proposed by Feigl and Rauch [4, 5]. They employed natural Gd and pure ^{157}Gd convertor foils to measure the escape probabilities and the energy distributions of the escaping electrons with Si surface barrier detectors.

Jeavons et al. [6] to apply a multiwire proportional chamber (MWPC) for the amplification and imaging of the secondary charged particles and discussed convertor materials other than Gd. To reduce the position broadening due to the large angular spread and range of the escaping fast electrons, a "high-density" drift space used as initial stage of the detector. This stage is essentially a multi-pinhole collimator where along each pinhole an electric field can established to drift of secondary ionization products towards to the amplification and imaging stage. However, the finite size of the pinholes limits the position resolution and the comparably large mass of the collimator increases the γ-ray sensitivity and reduces neutron detection efficiency.

In Melchart et al. [8] proposed a multistep avalanche chamber atmospheric pressure (MSAC) for imaging the electrons escaping from a Gd foil. The exponential avalanche growth in the parallel gap of preamplification step makes the detector more sensitive to charges produced near the entrance point of the particle track into the gas volume, thus ensuring a good localization of the neutron interaction in the thin convertor foil. Indeed, they achieved a position resolution better than 1 mm (FWHM).

We also have developed and made normal pressure MSAC with converter of natural gadolinium. During developing and operations we have made detailed studying of performances of MSAC. Possibilities stabilization of operation, influence of separate devices of detectors constructions on performances of detectors was studied. As a result, it was possible to refine considerably performances of detectors and to obtain long-term stability in-process detectors.

Breskin et al. [60] makes further step in development of multistep avalanche chambers they was development of low-pressure multistep avalanche chambers. In one of our publication, possibility of reception of a regime of a preamplification in the low-pressure multiwire proportional chambers also has shown at usage additional plain-parallel gap disposed on top of the MWPC [61].

Dangendorf et al. [62] has offered for the first time low pressure MSAC with gadolinium converter. It thus offered to use developing techniques to increase the low-energy secondary electron emission of the convertor foil. Despite the exponential growth of the avalanche in the first multiplication stage, the spatial resolution of neutron detectors based on low-pressure multistep avalanche chambers depends on the direction and range of the neutron-induced charged particles. Moreover, for weakly ionizing particles (like the conversion electrons from Gd) the detection efficiency depends on the probability of creating at least one slow ionization electron in the first few hundred micrometer thick gas layer of the preamplification gap. To increase the number of produced electrons, close to the convertor surface (secondary emitter) would improve the localization resolution and the detection efficiency was pleased.

Gebauer et al. [52] offered fast and high-resolution hybrid low-pressure micro-strip gas chamber (MSGC) detectors and being developed, which lend themselves for setting up large-area detector arrays.

Abbrescia et al. [63] offered Resistive Plate Chambers (RPCs) wide spread, cheap, easy-to-build and large size detectors. Here a technique, consisting in coating the inner surface of the bakelite electrodes with a mixture of linseed oil and Gd_2O_3 reported. This allows making RPCs sensitive also to thermal neutrons, making them suitable to employ for industrial, medical or de-mining applications. Thermal neutron-sensitive RPCs can operated at atmospheric pressure, are light weighted, have low γ-ray sensitivity and are easy to handle even when large areas have to be covered.

Takahashi et al. A new series of experimental imaging plate neutron detectors (IP-NDs) were made, where the composition of the respective IP-NDs, containing a photostimulable $BaFBr:Eu^{2+}$ phosphor and a neutron converter material, Gd_20_3 or ^6LiF, were varied systematically [64]. The different conversion processes between natural abundant Gd and ^6Li caused distinct imaging properties. The imaging steps and the factors governing the image quality are considered in the same way as X-ray radiography, and the quantum noise is estimated by the effective neutron absorption in these IP-NDs when they read by a BAS2000 image reader.

Gunji et al. [65] have been developing a new type of neutron imaging detector with position resolution better than 10 micrometer by absorbing liquid scintillator for neutron capture into a capillary plate by capillary phenomena. Establishing the methods to absorb liquid scintillator to the capillary plate and attaching it into a photomultiplier, we irradiated the detector with neutrons and alpha particles. From the results of basic experiments, it recognized that the capillary plate absorbing liquid scintillator could operate as a neutron detector.

Bruckner et al. [14] have been developing position sensitive silicon detectors and the corresponding electronics allow the construction of fast time response thermal neutron detectors. These detectors also exhibit excellent position resolution by combination of silicon detectors with thin Gd converter foils. Authors constructed several one- and two-dimensional prototype detectors. The position resolution and the detector efficiency for different converters at wavelengths from 1.1 to 3.3 Å were determined at the TRIGA reactor in Vienna and at the ILL in Grenoble. Spatial resolutions of less than 100 μm and efficiencies up to 40% have achieved. These detectors can also be used for phase topography experiments using perfect crystal neutron interferometers.

3.1 Neutron Imaging – Past, Present and Future (From the Site of NM13 [66])

Today, neutron imaging is at a pivotal point in its history. With the emergence of digital cameras and the rapid development of digital detectors with better spatial and temporal resolution, neutron imaging has developed into a valuable, reliable technique whose importance for materials research is finally being recognized. Many facilities around the world are planning to dedicate beam lines to neutron imaging or building imaging instruments. NMI3 recently started funding a Joint Research Activity in Neutron Imaging.

By Juliette Savin reviews the different neutron imaging methods available today, presents the new NMI3 Imaging Joint Research Activity and exposes some of the challenges still facing the technique.

Neutron imaging refers to a collection of non-destructive testing methods, which exploit the penetration of neutrons to investigate the internal structure of objects. According to the universal law of attenuation of radiation passing through matter, different materials attenuate neutrons differently. In a typical neutron radiography experiment, the object being investigated is placed in a well defined neutron beam. A 2D position-sensitive detector, often consisting of scintillator optically coupled to a CCD or EMCCD Camera, is placed on one side of the object and records the radiation transmitted through the object. The shadow image thus produced gives information about the internal structure of the object studied.

Whereas radiography results in a two-dimensional image of the sample studied, tomography allows to visualise samples in three dimensions. For neutron tomography, the sample rotates in the beam and multiple 2D radiography images are recorded. A 3D representation of the volume of the object can be reconstructed using a mathematical algorithm.

Few facilities in Europe offer access to instruments for radiography and tomography. At PSI´s NEUTRA, imaging is carried out thanks to SINQ´s beam of thermal neutrons while BEnSC´s CONRAD, FRM II´s ANTARES and PSI´s ICON offer cold neutron radiography and tomography. FRM II´s NECTAR makes use of the FRM II´s fission neutron spectrum. NRAD in Budapest uses BNC´s thermal neutrons.

Neutron imaging is complementary to other non-destructive imaging methods, in particular X-ray imaging. Whereas X-rays are scattered and absorbed by the electron cloud of an atom, neutrons interact with the atomic nuclei. Therefore neutrons are more sensitve than X-rays to light elements such as hydrogen, lithium, boron, carbon, and nitrogen. The degree of attenuation depends on the neutron energy. For some polycrystalline materials such as metals, a strong energy dependency of the attenuation is observed in the cold neutron range, due to Bragg scattering from the crystal lattice.

Although neutron imaging cannot compete with X ray imaging in terms of resolution, the high penetration depth of neutrons allows to study large samples of some materials, with volumes of up to 100 cm^3 at medium resolutions, which X rays cannot do. This can be extremely useful in fields such as engineering.

"When you need imaging, you will usually start with X rays, because they are much easier to get" explains Thomas Bücherl from TUM, Germany. "But it is a good idea to have both, because you get information with X-rays that you do not get with neutrons, and vice versa."

In some neutron imaging instruments, such as ANTARES at FRM II in Germany, there is the possibility to switch between neutron and X-ray beams to study the same samples.

In the early days of neutron imaging, the technique was done with film, using gadolinium as a converter. This allowed only for a certain type of science to be carried out with neutron imaging and many criticised the fact that the method wasn´t a quantitative method.

"Film is not a digital system, it is not linear, and it is very difficult to get quantitative information out of it. You need to scan the film point by point" explains Burkhard Schillinger responsible for ANTARES at FRM II in Germany. According to him, this is why neutron imaging didn´t really make it in the neutron world before 1990.

Since the early 1990´s however, neutron imaging has been transformed by the digital revolution and the use of CCD cameras instead of film. Suddenly, pixels could be counted, grey values could be quantified and imaging became a quantitative method.

Researchers then realised that they could get much better imaging, with higher resolution, by sacrificing intensity for higher collimation of their neutron beams – since the new electronic detectors were much more sensitive than film. And this has now become the key factor in neutron imaging instruments – collimators define the Collimation ratio, the ratio of the distance between the sample and the collimator aperture over the diameter of the aperture. The intensity of the neutron beam inevitably drops with collimation, but this can be compensated for by using very sensitive detectors.

The unique ability of neutrons to easily penetrate metals and still be sensitive to light elements such as hydrogen, carbon, and nitrogen makes neutron imaging particularly useful for applications in fundamental research and industry.

For instance, neutron imaging is used routinely to highlight light materials such as hydrogenous substances with high contrast in engine parts or in hydrogen storing tanks and fuel cells.

Figure 44. Radiograph of an analog camera: by neutrons (top) by X-rays (bottom). While X-rays are attenuated more effectively by heavier materials like metals, neutrons make it possible to image some light materials such as hydrogenous substances with high contrast: in the X-ray image, the metal parts of the photo apparatus are seen clearly, while the neutron radiograph shows details of the plastic parts.

Neutron imaging also allows to visualise the movement of fluids, such as oil or water, in large metal objects.

Neutron imaging has attracted the attention of companies in the car and aeronautical industries, which regularly visit neutron imaging facilities to carry out quality control tests or studies on engines, gear boxes or other metal parts.

Neutron imaging is non-destructive in nature, which also makes it an attractive method to analyse archaeological artefacts or pieces of Art.

Another application of imaging is the study of wooden objects, both for engineering or archaeological purposes.

Despite the wealth of applications of neutron imaging, neutron facilities have a hard time attracting users from the industry or archaeology worlds.

"Although they produce nice pictures, which can be attractive, our methods often seem too exotic to companies or museums" explains Eberhard Lehmann, Imaging group leader at PSI in Switzerland.

Users from outside who come to neutron facilities to do imaging often have problems interpreting the results they get with neutron imaging. "Very often users just leave the interpretation down to us, because they really don´t know what to do with their results" explains Anders Kästner, expert in imaging and ICON instrument responsible at PSI.

Despite the broad range of applications of neutron imaging, neutron imaging instruments have historically been given little space or importance within neutron facilities. "In a typical neutron facility, you´ll often find only one neutron imaging instrument" explains Dr Lehmann. And neutron imaging instruments need space, which is not always available at existing facilities to build new imaging instruments.

"For a long time neutron imaging wasn´t considered science. People thought: You´re just producing coloured images, that´s not science!" explains Burkhard Schillinger. Facilities didn´t want to dedicate beam lines to neutron imaging. But the digital era has dramatically transformed imaging and neutron imaging is experiencing a sort of new beginning, with facilities around the world suddenly investing in neutron imaging instruments.

According to Eberhard Lehmann, "there are indeed some good indications for an increase of the neutron imaging activities world-wide".

Indeed, a neutron imaging and diffraction instrument for materials science, materials processing and engineering, IMAT, is currently in design phase at ISIS in the UK. In Norway, Kjeller is considering dedicating a beam line to neutron imaging, and one is under construction at IBR-2 in Dubna, Russia. In the Czech Republic, a test beam line is available at the Rez reactor. There are of course plans to build a neutron imaging facility at the ESS and the ILL in Grenoble is considering building a new neutron imaging facility.

Further afield, the reactor CARR in China already possesses two dedicated beam lines and ANSTO in Australia, J-PARC in JAPAN and SNS in the USA are planning their first neutron imaging facilities.

"These are really positive approaches and activities into the right direction" confirms Dr Lehmann, who is also president of the international Neutron Radiography society. All this suggests that the international neutron imaging community, which at the moment comprises around 200 people, excluding one-off users, is bound to grow. The community meets every two years alternately at the International Topical meeting on Neutron Radiography, which this year is taking place in Canada, and the World Conference on Neutron Radiography, which will be held in Switzerland in 2014.

New, more sophisticated neutron imaging techniques have been developed recently, which also explains the new interest in neutron imaging. Techniques such as real-time imaging, stroboscopic imaging, energy selective imaging or neutron grating interferometry are now carried out routinely at facilities across Europe. They extend the scope of neutron imaging beyond conventional attenuation imaging and provide information about properties and phenomena until now difficult to study.

The development of highly efficient detector systems allows for real-time imaging, which can be carried out at most neutron imaging facilities with an intensive beam. This technique allows to follow a dynamic process in real-time. The limited brilliance of today´s neutron sources only allows exposure times of a few milliseconds per image, but better time resolution can now be obtained with stroboscopic imaging.

The first stroboscopic imaging experiment was carried out in 2002 in collaboration between FRM II, the ILL, the University of Heidelberg and PSI. A four-piston BMW combustion engine was inspected by stroboscopic neutron radiography technique at the high flux test beam line H9 at ILL achieving exposure times of few hundred microseconds. Since then, stroboscopic imaging has proved to be ideal to visualize the dynamics of lubrification liquids and cooling agents in metal casing and has become the method of choice for the imaging of combustion engines in operation.

Neutron imaging can provide more accurate information on a sample if carried out with selected neutron energies. The proof of principle experiment for energy-selective imaging was carried out at ISIS. Use of this method requires a cold neutron spectrum. In this method, the transmission signal is recorded position sensitively depending on the neutron energy being varied. Due to Bragg scattering in the crystal lattice of the material in the sample, a diffraction contrast is observed in the transmission images. This contrast allows to distinguish between different crystallographic modifications of the material and hence to make phase separation in 3D, non-destructively. The diffraction contrast provides information about the presence of mechanical residual stresses and textures in the sample.

Neutron grating interferometry also requires a cold spectrum and it is not available everywhere. In Europe, two instruments provide this method as option to the users. ICON, the cold neutron imaging instrument at PSI, is one of the only instruments in Europe to offer energy selective imaging or neutron grating interferometry. A grating interferometric setup was recently was installed on the CONRAD beam line at HZB, in collaboration with PSI. Another grating interferometer will be installed at the upgraded ANTARES facility at FRM II soon. Grating interferomery can be used to investigate micro- and nano-heterogeneity of structures with sizes ranging from 0.1 to 10 μm.

Imaging of nano and magnetic structures. Better neutron imaging technology is still needed for imaging of very small structures and to improve the scope of magnetic imaging. Good, high-resolution magnetic imaging is crucial in particular for the study of superconducting materials.

NMI3 has decided to act on this by putting together a Joint Research Activity (JRA) focusing on the development of better components for neutron imaging and of new techniques using different interaction mechanisms of neutrons with matter to produce contrast, than conventional absorption imaging methods.

The JRA, funded by NMI3 and involving teams from FRMII, PSI, HZB, ISIS, LLB, JCNS and NPI, will focus in part on combining neutron imaging and scattering techniques, such as grating interferometry, which can be used to investigate structures at the micro and nano level together with innovative imaging with polarised neutrons, which can visualise magnetic fields in the direct space. The goal of this project will be the visualisation of microstructural changes in metallurgical samples under tensile or torsion stresses.

Another task within this JRA will be to develop techniques for vectorial magnetic imaging in the volume of nano and micro-structures. Existing magnetic imaging techniques (X-ray Photoemission Electron Microscopy, Atomic Force Miscroscopy, electron holography) are mostly surface techniques and do not provide direct information about the volume magnetisation of nano-objects.

One of the main limitations in neutron imaging right now is the spatial resolution of the detectors available today. As part of the JRA, a team at FRM II will also work on the development of better detector components, for better high-resolution imaging.

The other limitation is the neutron flux available for experiments. "We should optimise our detectors and also our beams for neutron imaging experiments. At the moment we cannot compete with synchrotron radiation facilities, which have better flux and better intensities" explains Nikolay Kardjilov, coordinator of the Imaging JRA.

One way to optimise neutron beams is to improve optical components used to focus the beams, the aim of another Joint Research Activity supported by NMI3, the Neutron Optics JRA.

The future is looking bright for neutron imaging. If it wants to compete with other imaging techniques, more available, easier to carry out and to interpret, such as X rays, effort and funding for technical development will have to continue to flow.

3.2. Normal Pressure Multistep Avalanche Chamber

A multistep avalanche chambers (MSAC) consists of a conventional multiwire proportional chamber (MWPC) placed within a single gaseous volume with additional grid electrodes forming preamplification and drift spacing. Gadolinium converter settles down directly ahead of a preamplification grid. MSAC registers electrons escaping off in a back hemisphere. In the preamplification gap, there forms an electron-photon avalanche, which transferred to MWPC for terminal amplification and registration. Due to the exponential amplification in the preamplification gap there is implemented a tie of coordinates to the entrance point of particles to the detector volume. The spatial resolution of MSAC can be improved by the amplitude analysis and the consequent mathematical processing of events up to 0,2-0,3 μm. The gas amplification coefficient of MSAC is of the order 10^{6-7}, which provides registration of single electrons with practically zero energy. The registration efficiency of the MSAC-based detectors mainly defined of converter's characteristics.

In the Physical Technical Institute of Academy of Sciences of Republic Tajikistan, the detector based on MSAC with natural Gadolinium converter is developed and created. Appearance of the detector showed in Figure 45.

The basic parameters of detectors are: sensitive volume - 200x200 mm, gaseous mixture - Ar + (1,5%) of n-geptane, operation pressure - 1 atm, detection efficiency (at λ = 1.8 Å)-15%, spatial resolution-<1mm (FWHM), time resolution (by anode signal)- 20ns, counting rate-10^6 1/s.

Figure 45. Appearance of the detector and serving electronics.

Studying of performances of detectors makes with use gamma- and beta-sources. Especially well for these purposes approaches β - a source ^{14}C with the maximum energy of electrons 150 keV. Taking into account energy weakening of electrons in an input window of the detector the energy distribution of electrons becomes closest to an energy distribution of electrons escaping from the gadolinium converters at response of a radiative capture of neutrons by gadolinium nuclear.

3.2.1. Characteristics of the Multistep Avalanche Chambers

Structurally MSAC consists of MWPC, located in common gas volume with electrodes A, B and C (Figure 46) on which corresponding potentials brought. These electrodes form conversion (AB), preamplification (BC) and the drift (CD) gaps. In gadolinium converter electrons generated, and only their part can escape a converter material to the conversion gap. In conversion, gap formation of the secondary electrons under the influence of a feeble electric field (~ 2 kV/cm) is carried out. Generated electrons transferred in to plain - parallel preamplification gap. In this gap in the strong electric field (up to 10 kV/cm) the electron-photon avalanche develops. Only part of electrons, from the formed avalanche, can reach the drift gap and transfer to multiwire proportional chamber. In MWPC its further amplification and registration carried out.

The basic contribution to avalanche development in MSAC is a photo - ionization process. That is, ionization of additive atoms by the quantum which are let out by atoms of rare gases at the transition to the basic state.

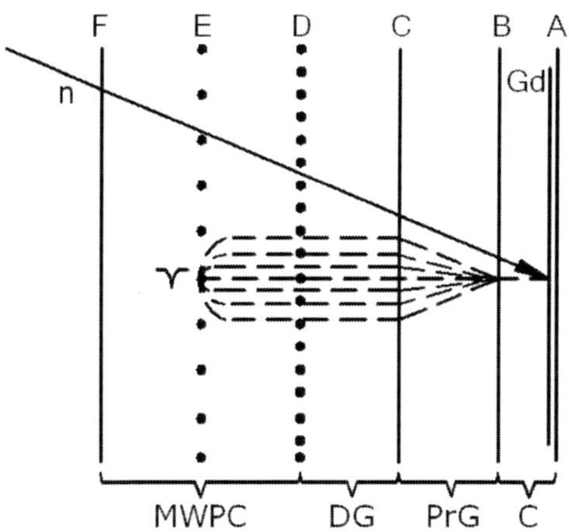

Figure 46. Scheme drawing of the detector, where MWPC-multiwire proportional chamber, DG-drift gap, PrG- preamplification gap, C- conversion gap, Gd- gadolinium foil.

The quantity of the molecular additives in gas mixtures MSAC usually is composed by 1- 4%. At higher concentrations, despite reduction of a free length of secondary electrons the avalanche develops less effectively. It, closely associated with reduction of medial velocity of drift of electrons and, accordingly, with reduction of their energy.

Rare gases and usually argon as basic components in gas mixtures are used. The photo - ionization process of amplification occurs on mixtures of argon with acetone, a bensole, n-geptan, etc. The coefficient of gas amplification MSAC is represented expression:

$$G = k \cdot Gpr \cdot G_{MWPC}, \tag{47}$$

where: Gpr - and G_{MWPC} - coefficients of a gas amplification of a preamplification gap and MWPC. The coefficient (k) characterizes efficiency of transfer of an avalanche from gap with high electric strength field (a preamplification gap) to the gap with low field strength (the drift gap) and is equal to ~Edr/Epr. The relation of fields in the chamber makes 1/5- 1/10 and only 10- 20% of electrons transferred in to drift gap, the quantity of transferred electrons less than the specified value is real and depends on an optical transparency of electrodes. Strength of electric fields in the drift gap and in MWPC it is approximately identical that allows, practically free lost to transfer electrons from the drift gap in MWPC.

Avalanche development process has a purely exponential character. In preamplification gap in strong (up to 10 kV/cm) electric field, drifting electrons form the electron -ion steams. Generated electrons in turn participate in the further development of an avalanche.

The gas amplification coefficient described by expression [7]

$$Gpr = n/n_0 = \exp(\alpha L), \tag{48}$$

where, α – first coefficient of Tausend, L - a space of a preamplification gap. This expression is valid for a charge formed in the upper stratum of the gap of preamplification. At registration quantum their absorption occurs along all gap and the amplification gain for each case will describe as

$$Gpr = \exp(\alpha (L - Xi), \tag{49}$$

where: Xi - depth of absorption quantum along a gap.

At registration of the charged particles primary ionization radiation occurs on all trajectory of a particle and the gas amplification coefficient in this case describe as

$$Gpr = (\exp(\alpha L) - 1)/(1 - \alpha L), \tag{50}$$

The coefficient of preamplification gain of MSAC can reach quantities 10^5 -10^6, and the full gain coefficient G = 10^7. At achievement the critical sizes ($\alpha L = 20$) of avalanche, corresponding to quantity of electrons in an avalanche ~ 10^8, the avalanche according to a rule of Reter develops into a streamer. The streamer leads to a spark disruption in MSAC. This rule enters restrictions on greatest possible Gpr. Therefore, at registration of one-electron events Gpr can make quantity 10^6 -10^7. At registration of events with big specific ionization losses quantity maximum achievable Gpr decreases, for example Gpr - 10^4 -10^5 at registration gamma -quanta with energy of 6 KeV and ~ 10^2 at registration α - particles with energy in some MeV.

3.2.2. Gas Amplification and Efficiency of Registrations MSAC

Measuring of coefficient of preamplification gain (Gpr) for various gas mixtures spend. Gpr it was measured as the relation of an amplitude of signals from an anode plane at change of a high-voltage on a preamplification gap. Coefficient Gpr can reach quantities 10^5, at Gpr> 10^4 rare spark disruptions in a preamplification gap observed. With magnification of an amplification values intensity of sparking is incremented. At Gpr> 10^5 the continuous spark discharge localized under a radiation sources is observed.

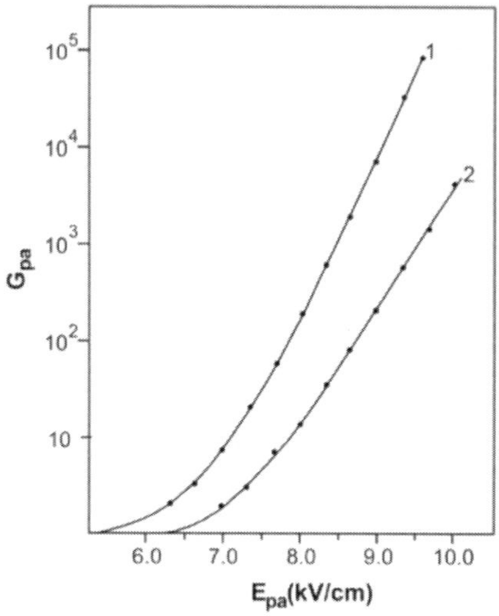

Figure 47. Factor of preliminary amplification in dependence on high voltage on pre-amplification gap. A gas mix of argon with 1,6% of acetone (a curve 1) and 3% of acetone (a curve 2).

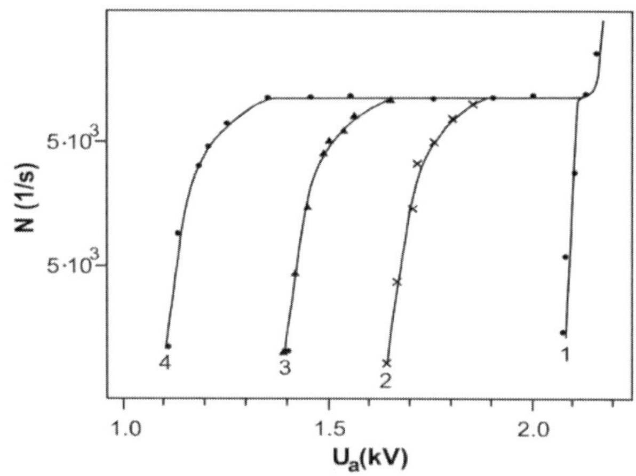

Figure 48. Counting characteristics of the detector at registration γ-quanta ^{55}Fe on dependence on high voltage on MWPC. Value Epr is equal 2 kV/cm (1); 8,3 kV/cm (2); 9,0 kV/cm (3) and 9,7 kV/cm (4).

Value of coefficient of preamplification gain can reach 10^5, at registration γ-quanta with energy of 6 KeV. It is necessary to note that at Gpr> 10^4 rare spark disruptions on the area in a preamplification gap observed. At definition of the full coefficient of preamplification gain in a gap, it is necessary to consider losses of electrons while transferring avalanches in the drift gap. For a relation of fields of the drift and preamplification gap 1/5 full value Gpr = $5*10^5$ that corresponds to value of coefficient $\alpha = 4{,}4\ mm^{-1}$. Common amplification gain of MSAC taking into account amplification in MWPC can reach values G = 10^6 -10^7. At Gpr > 10^6 the recurring operation, MSAC caused by a photon feedback observed. A part of the ultraviolet quanta formed in MWPC, can reach a Gd converter, causing occurrence of photoelectrons from materials, impulses from which observed through 1μs after the first signal. At Gpr> 10^7 avalanche develops into a streamer that comes to a spark disruption in MSAC.

Examinations efficiency registration and the time resolution spent on a gas intermixture of argon with 2% of acetone. Extent of a plateau counting rates studied at registration γ- quanta ^{55}Fe. In Figure 48 given extends of plateau counting rates of signals from anode MWPC at various values electric field strength of in a preamplification gap depending on electric voltage on the anode MWPC (Ua). The curve 1 in Figure 48 measure at low intensity of Epr, thus an avalanche did not develop and only transfer electrons from conversion to the drift gap, carried out. At Epr = 8- 10 kV/cm quantity Gpr essentially increases that considerably increments extent of a plateau counting rates (curves 2-4). It is visible that the plateau reaches quantities 800 V in at value Gpr 10^4.

Efficiency of registration of charged particles with the minimum ionization losses was explored with β - particles ^{90}St (Emax = 0,53 MeV). In Figure 49(A) observed dates depending on sense of anode voltage MWPC for the various field gradients in a preamplification gap presented. Measuring spent in the case of absence of a conversion gap. At enough high values of Epr with efficiency ~ 98% register the detector β - particles, quantity of a plateau makes 300 V.

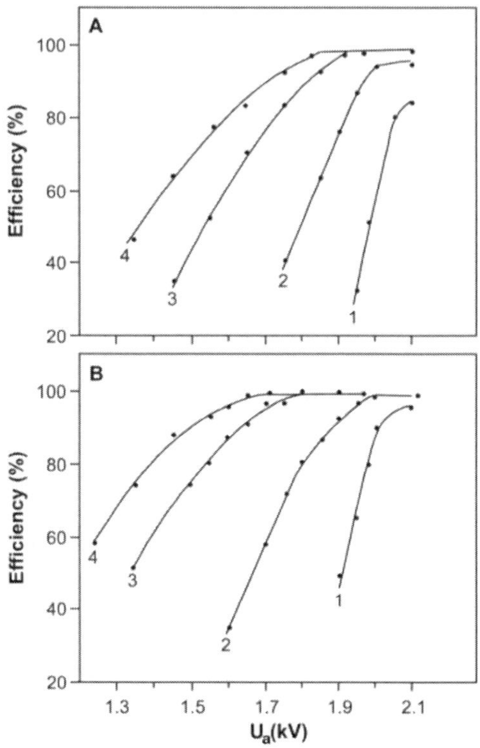

Figure 49. Efficiency of the detector at registration β- radiations depending on voltage on MWPC. Value Epr is equal 8,3 kV/cm (1); 9,0 kV/cm (2); 9,7 kV/cm (3); 10 kV/cm (4). A - the detector without conversion gap; B - size of conversion gap AB = 1 mm.

As it told above, quantity of Gpr for the secondary ionization electrons formed in different points on a thickness of a preamplification gap is various. Introduction of a conversion gap by 1 mm leads to extension of a plateau of efficiency of the detector and allows reducing value of necessary anode voltages.

3.2.3. Spatial Resolution of the Multistep Avalanche Chambers

Multistep avalanche chambers (MSACs), as well as plane-parallel avalanche detectors, as distinct from the ordinary multiwire proportional chambers (MWPCs), have high accuracy in determining coordinates along both the X – and Y- axes, the presence of the so-called "focusing effect" (i.e., the coordinates of the point at which a particle hits the detector's sensitive volume are measured) and a high gas amplification coefficient.

An electron–photon avalanche that develops in the preamplification gap transferred through the drift gap to the MWPC for further amplification and detection. Coordinate data read out from the MWPC cathodes via delay lines, and the anode signal can used for amplitude selection and triggering of events.

The path of photoelectrons, Auger electrons, and fluorescent quanta's limit the spatial resolution of gas-filled position-sensitive detectors (GFPSD) for a narrow collimated γ-ray beam. For example, 5.9-keV γ-rays are converted in argon, mainly on its K subshell with a 3.2-keV ionization potential escaping photoelectrons with an energy of 2.7 keV have a range of 30 $\mu g/cm^2$ (i.e., 250 μm of argon at normal pressure and temperature). A fluorescent quantum is produced in 88% of events; its energy is 3.2 keV and its mean free path in argon is ~40 mm. Absorption of a fluorescent quantum inside the detector results in emission of an electron with an energy of 3 keV and a range of 250 μm.

Let us consider the effect that the photoelectron range exerts on the spatial resolution of a GFPSD. Conversion of γ- rays assumed to occur at one point, and detection efficiency for fluorescent quanta thought to be low and can ignored. Photoelectrons escaping from the conversion point in all directions form a spherical cloud of secondary electrons with radius Re equal to the photoelectron range. The full width at half-maximum (FWHM) of the projection of this cloud onto the plane of the detector is 3/2 R_e. The dependence of the spatial resolution on the electron energy can also found from the empirical formula in [67]:

$$\sigma\ (\mu m) = a\ E_e\, n/\rho, \qquad (51)$$

Where E_e [keV] is the electron energy; ρ [mg/cm2] is the specific density of the detecting medium; and the coefficients have the following values:

a = 30, n = 1.30 for E_e = 1–5 keV and

a = 17, n = 1.64 for E_e = 5–20 keV.

Apart from the photoelectron range, the spatial resolution is dependent on other factors too, particularly on the resolutions of the delay line and the data acquisition electronics.

The spatial resolution of the MSAC investigated by using a highly collimated (slit width, 50 μm) ^{55}Fe source, which was moved over the detector surface. The experimental results presented in Figure 50. The FWHM was 400 μm. Taking into account the contributions of R_e and the collimator, the intrinsic resolution of the detector was 260 μm (FWHM) [68].

When MWPC detects charged particles from an isotropic emitter, such as Gd converters, its spatial resolution is rather poor. This explained by the fact that, in the case of oblique incidence, secondary electrons produced along the track of electrons form an extended cloud of charges that collected onto the MWPC anode plane and the center of this cloud is displaced from the particle's hit point through at significant distance. As distinct from the conventional MWPC, gas amplification in the pre-amplification gap of the MSAC obeys to the exponential law; therefore, the first electrons produced in the upper layer of the gap were responsible for ~80% of the whole signal amplitude. This improves substantially the spatial resolution of the MSAC during detection of isotropic radiation and gives rise to the so-called "collimating" effect of the MSAC.

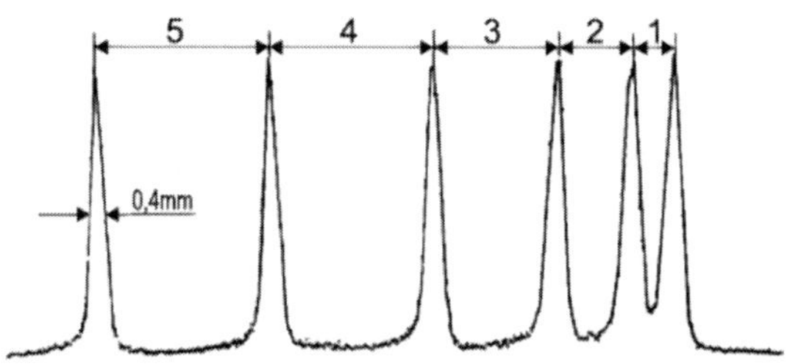

Figure 50. Spatial resolution of the MSAC. The highly collimated ^{55}Fe source moved with a step of 5, 4, 3, 2 and 1 mm.

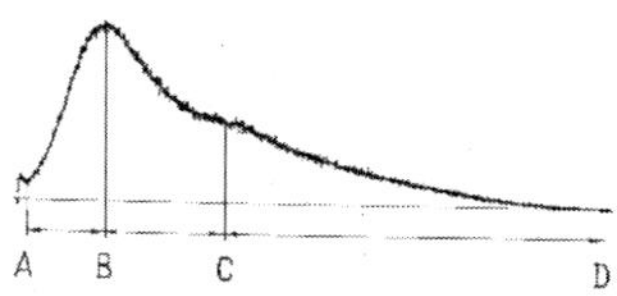

Figure 51. Amplitude spectrum of β particles from the ^{14}C source.

In the first operations on studying of performances of MSAC with gadolinium converter [8] dependence of the space resolution on amplitude of displayed signals has detected. The amplitude of a signal is depending from angle of an escaping electrons from a converter material.

The spatial resolution can improve by using amplitude selection of events in electron detection [68]. With this aim, we used two radiation sources (^{14}C) with approximately equal activities, shaped as spots ~1mm in diameter and spaced 1 mm apart (the spacing between the boundaries of these spots). The sources placed on the entrance window of the detector. Anode signals used for amplitude selection of events by differential amplitude discriminator. The gating signal for data acquisition generated by the discriminator and arrived at the control input of the time-to-amplitude converter. Information acquired with a multichannel amplitude analyzer. The amplitude spectrum of the anode signals from the detector irradiated by β-particles of an uncollimated source (^{14}C) shown in Figure 51.

Particles incident on the surface of the chamber at small angles travel a significant distance in the upper layers of the pre-amplification gap, giving rise to additional avalanches due to secondary electrons along primary particle tracks. These particles are associated with higher-amplitude signals. As a result, as the angle of incidence decreases, the signal amplitude increases, the electron cloud grows in size, and its centroid displaced toward the direction of the primary particle. A small bend on the right of the spectrum can be attributed to the presence of a 0.5-mm-wide conversion gap and represented by the superposition of two spectra, one of which is formed by particles that suffered interaction in the conversion gap and generated secondary electrons and the other of which is due to particles that passed the conversion gap without interaction.

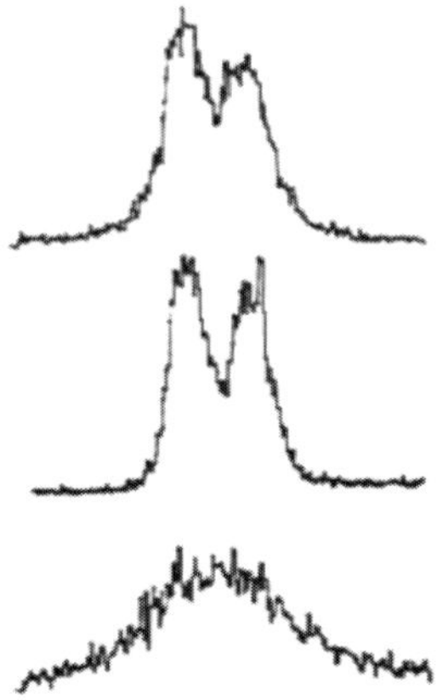

Figure 52. Spatial resolution of two radioactive zones of ^{14}C isotope, each 1 mm in diameter, located with 1-mm spacing between the boundaries.

The coordinate resolution obtained when all events were detected (segment AD of the curve in Figure 51) shown in Figure 52 (curve 1). The ratio of the depth of the dip between the peaks to their height is 0.25. It is possible to improve the spatial resolution by selecting (using the differential discriminator) events corresponding to the portion of the anode signal spectrum from A to B (Figure 51). Incidentally, the ratio of the depth of the dip to the peak height is 0.4 (curve 2 in Figure 52). In this case, the spatial resolution limited by the own resolution of the detector system. Which is dependent on both, the delay time per unit length of the delay line, and on the accuracy of the electronic channels. When the coordinate information obtained from events with high signal amplitudes (segment CD in Figure 51), the spots cannot be resolved (curve 3 in Figure 52).

The spatial resolution of the MSACs has investigated. The intrinsic spatial resolution is 260 μm (FWHM) or 100 μm (σ). The spatial resolution of the detector under exposure to soft γ-rays of the ^{55}Fe source is 400 μm (FWHM). The spatial resolution of the MSACs is dependent on the angle of incidence of conversion electrons. The high signal amplitudes due to particles incident on the surface of the chamber at small angles (25% of all events) considerably impair the spatial resolution. By analyzing the signal amplitude and selectively measuring the coordinates, or by estimating the signal amplitude and its contribution to the reconstructed image, it will be possible to raise the spatial resolution when electrons escaping from solid-state neutron (Gd) or γ-ray converters detected.

3.2.4. Influence of Gas Mixes on Characteristic MSAC

Researches of influence of additive concentration on characteristics of the MSAC brought on a gas mix argon + acetone. Acetone as the additive has chosen from - that allowed changing in a wide range concentration of the additive in a mix. In Figure 53 values of dependence of factor of preliminary amplification (Gpr) from value of an applied voltage (Upr) for various values of concentration of acetone are brought. From figure, it is visible, that with increase of concentration of acetone at fixed Upr the factor of preliminary amplification decreases. Thus, Gpr can increase at corresponding increase Upr. Reduction of factor Gpr is connected with decreasing of average energy of drifting electrons as dominating process in interaction electrons with molecules of the additives. This interaction leads to sharp loss of energy of electrons. Reduction of average energy of electrons lead to decreasing energy interaction with molecules of additives. Then excitation of molecules occur without radiation via rotary and oscillatory transitions. Dependence of amplification on concentration of the additive at the fixed voltage on a preamplifying gap well described by following expression [69];

$$G\,(p) = 10^{-(n \cdot pi)}, \tag{52}$$

where: pi - concentration of the additive in%: n - the factor depending from pi and for a range 0,5 ÷ 3% is equal 2. So change pi within the limits of ± 0,5% cause change G in ± 10 times. The dotted curve designates border of sparking in a pre-amplification gap. Optimum for achievement of the maximal factor of amplification are concentration 1,5 ÷ 2%.

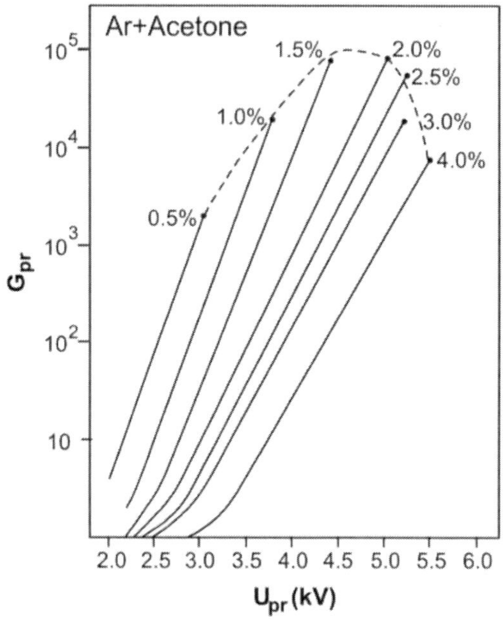

Figure 53. Dependence of the factor of preliminary amplification MSAC for various values of concentration of acetone in argon.

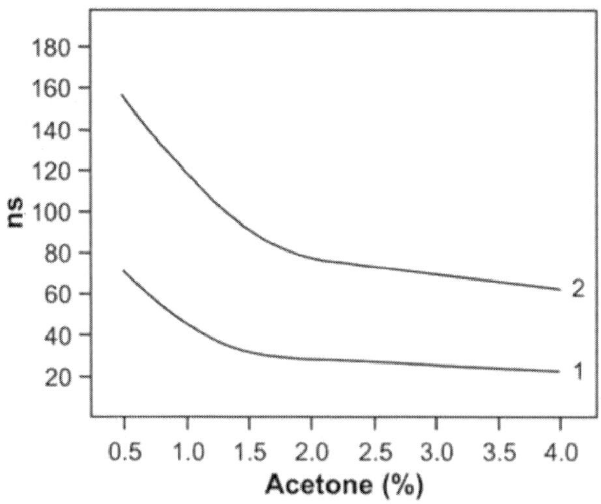

Figure 54. Dependence of time resolution of MSAC on concentration of acetone in a gas mix. A curve 1 - FWHM, a curve 2 - width at a level of 97% of the registered events (FWTM).

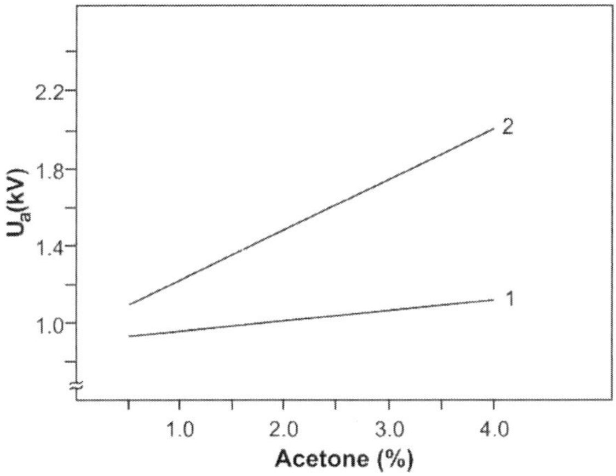

Figure 55. Extent of a plateau counting characteristics of MSAC for various concentration of acetone in a mixture. Curves 1 and 2 accordingly, the beginning and the end of a plateau for a source ^{55}Fe.

Influence of concentration of the additive on time characteristics MSAC investigated by β - a source 90St. Measurements were spent at fixed values G and Gpr. In Figure 54 dependences of the time resolution of the detector for a mix argon + acetone are resulted at various concentration of acetone. The curve 1 shows FWMH, a curve of distribution of 2 widths at a level of 97% of the registered events FWTH. From figure it is visible, that with increase of concentration of the additive time resolution of MSAC improved, Time resolution for concentration of additive -4% makes 21 nanoseconds (FWMH) and 60 nanoseconds at a level of 97% of the registered events.

Researches of influence of acetone concentration on length of plateau counting characteristics carried out. Researches were spent at fixed factors Gpr. The curve 1 in Figure 55 shows voltage on anode MWPC, corresponding to the beginning of a plateau for a source ^{55}Fe. The curve 2 shows voltage MWPC corresponding to the end of a plateau counting characteristic. From figure it is visible, that with increase of concentration of the additive extending of a plateau occur. Basically, its occurs due to displacement of the end of a plateau. Therefore, extent of a plateau makes 900 V for the concentration of acetone equal 4%.

3.2.5. Ways of Stabilization of Operating Modes of MSAC

Proceeding from above the characteristic show, that the highest factor of amplification, can be reached for a mix of argon with 1,5 ÷ 2,5% of the additive. At the same time, the factor of amplification strongly depends on variation of concentration of the additive (in ± 10 times) at change of concentration of the additive on ± 0,5%. Other characteristics of MSAC depend strongly on the variation of concentration of the additive too. In order to receive the long-term stable operating mode of MSAC it is necessary to stabilize concentration of organic compound applied as additives.

One of possible ways of stabilization of concentration of additives is thermostatic control of flasks with a liquid through which gas passed. Thermostating can carry out by means of thermostats, or at the temperature of a thawing ice. For maintenance of concentration of known additives within the limits of 1 ÷ 3% it is necessary to pass part of gas through a flask with a liquid. It dictates necessity of stabilization of gas speeds by means of highly stable reducers. However, even at use of similar reducers there necessity to expose and supervise carefully of gas blowing by rotameters. For the further increase of stability of work of MSAC it is necessary to use new gas mixes.

The most convenient in operation are organic compounds with low-pressure of saturated vapor, which at passage through them of all argon at temperature 0°C allow receiving concentration of the additive in a range 1 ÷ 3%. Such mixes are argon + H-heptanes and argon + isopropyl spirit. At passage of all argon through H-heptanes and isopropyl spirit at 0°C their concentration, accordingly makes 1.5 and 1.1%. Received data on these mixes are similar to the characteristics on mixes argon + acetone.

Other interesting gas mixes for work MSAC are gas mixes on the basis of neon which emitted more vigorous (in comparison with argon) photons at transition in the basic condition. Energy of photons is sufficient for ionization of some organic additives, in particular methane.

The gas mixes neon + methane is convenient at operation as allows preparing binary mixes in high-pressure tanks. Researcher of characteristics MSAC on this gas mix carried out. The mix allows to receive high enough factor of preliminary amplification Gpr = 10^4. However, at the same time a neon + the methane mix does not allow receiving a wide plateau. At registration of relativistic particles, extent of a plateau is 100 V for concentration of 2.8% of methane. The plateau of efficiency of registration of relativistic particles can increased at addition of argon to the neon mixes. Therefore, extent of a plateau of efficiency increases by 100 V for gas mixes Ne + (10 ÷ 15%) Ar+ methane. Neon mixes has more higher (2 times) the time resolution in comparison with argon. So the time resolution for mix Ne ÷ of 2,2% of methane makes 15 nanoseconds (FWHM). Mixes based on neon can applied to detecting low vigorous electrons in conditions raised gamma – background, first in the neutron detectors with gadolinium convertors.

3.2.6. Detector Design

Physically, each MSAC consists of a series of wire electrodes located one after another in a common gas volume, thereby forming conversion, preamplification, and drift gaps and a MWPC. The schematic diagram of the MSAC it's shown in Figure 56.

The conversion gap formed by convertor window *A*, which made from gadolinium foil and electrode B. The preamplification gap confined between grid electrodes B and C (these are commercially produced grids or grids composed by orthogonally wind wires). It should be noted that efficient transfer of an electron–photon avalanche from the preamplification gap to the drift gap can be obtained only when the grids have a high optical transparency.

In [70], it noted that stable operation of the MSAC could obtain only when the grids made from thick wires (50–100μm in diameter). For grids composed from 20-μm diameter wires, sparking in the gap observe it make impossible to achieve preamplification coefficients *E*pr> 10. Because drift electrode C begin playing the role of a quasi-anode and development of an electron–photon avalanche near the wires becomes a dominant process of this electrode, which increases significantly the probability of a spark breakdown in the preamplification gap.

The wire pitch in the grid is governed by the requirement that the electrostatic field in the preamplification gap be uniform. The electrostatic field assumed uniform at distance $L \geq 3\lambda$, where λ is the grid period. Taking into account, that two grids (B and C) form the preamplification gap, it is desirable that the wire pitch selected be $\lambda \leq L/6$, where L is the width of the preamplification gap.

The width of a preamplification gap usually selected from a range of 3–8 mm. However, the operability of an MSAC with a 20-mm preamplification gap was demonstrated in [71]. The preamplification coefficient Gpr of the MSAC is exponentially dependent on distance (L) and the first Townsend coefficient (α),

$$G\text{pr} = (e^{\alpha L} - 1)/(1 - \alpha L), \quad (53)$$

and increases with decreasing gap width. However, in the case of inclined tracks, the preamplification gap width should reduce to achieve a high spatial resolution. Incidentally, the requirements for the parallelism of the gaps become more stringent.

Stable operation of an MSAC can be achieved without using the drift gap. We investigated the characteristics of such a two-step structure. Two-step MSACs also exhibit high gas amplification coefficients, but are not free from some drawbacks. In the case of a spark breakdown, the spark short-circuits the cathode plane, which in turn may lead to a local breakdown of the delay lines and damage the amplifiers.

Creation of the uniform electrostatic field in the preamplification gap is fraught with difficulties due to the cathode discontinuity. MSACs with the inductive readout permit the obtaining of equal spatial accuracies for both coordinates. However, for coordinates along the *Y-axis* (parallel to the anode wires), this holds true if an electron avalanche covers several adjacent wires, forming cluster events. The avalanche expands occurs in the preamplification gap in the process of its development and during its transfer through the drift gap in the MWPC.

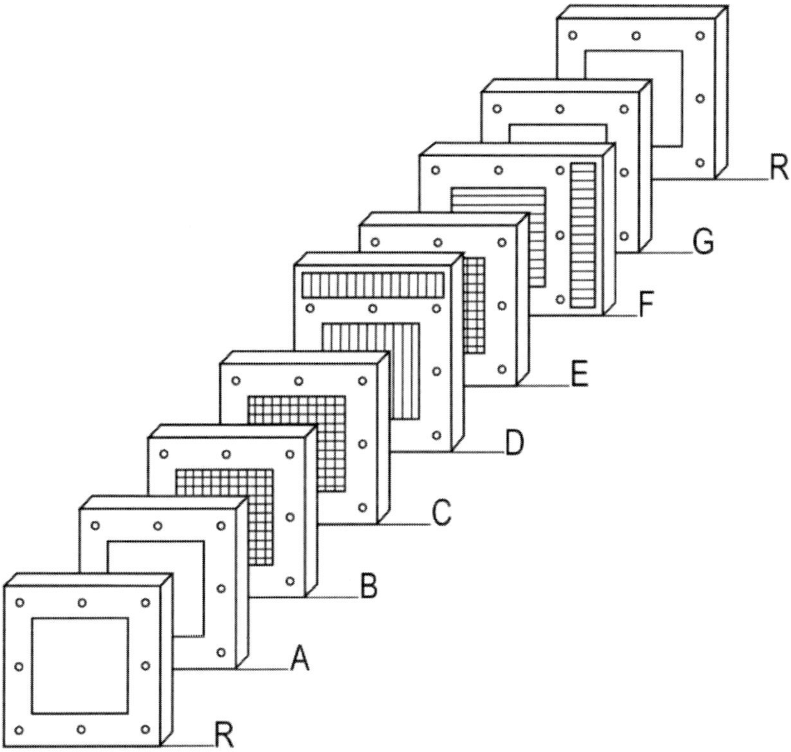

Figure 56. Schematic diagram of the MSAC.

As a rule, the widths of the drift gap are 5–10 mm. The use of wider gaps requires that a higher voltage be applied to the drift and preamplification gaps and gas volume of detector was increment.

A MSAC usually comprises a standard MWPC composed of an anode plane and two orthogonal planes X and *Y*. The anode plane is formed by a gold-plated tungsten wire 20μm in diameter, positioning with a pitch of 2 mm. All the wires combined into a common busbar. The cathode planes formed by beryllium- bronze wires, 50–100 μm in diameter, positioning with a pitch of 1 mm and combine into the strips. Each strip comprises four wires and soldered to the delay line. The standard width of the anode-to-cathode gaps is 5 mm.

3.2.7. Effects of Some Constructional Elements on the MSAC Characteristics

The uniformity of gas amplification coefficient *G* over the detector area is an important characteristic of an MSAC. A difference in the values of *G* can lead to nonuniformity of the detection efficiency. Figure 57 shows the distribution of *G* over the MSAC area [54].

Taking the logarithm of Eq. (53), one can obtain deviation of the preamplification gap width ΔL versus the relative variation in the MSAC amplification coefficient:

$$\Delta L = (1/\alpha)/(\Delta G/G) = (L/\ln G)/(\Delta G/G), \quad (54)$$

where: G is the mean average of amplification coefficient over the chamber area. This formula helps estimate the deviations from the parallelism in the preamplification gap.

The deviations of the gap width in the MSAC detector from an ideal plane (dashed lines), which calculated according to Eq. (54), shown in Figure 58. The deviations of mean amplification *G* are actually dependent both on the parallelism of the preamplification gap and on the spread of the MWPC amplification coefficient.

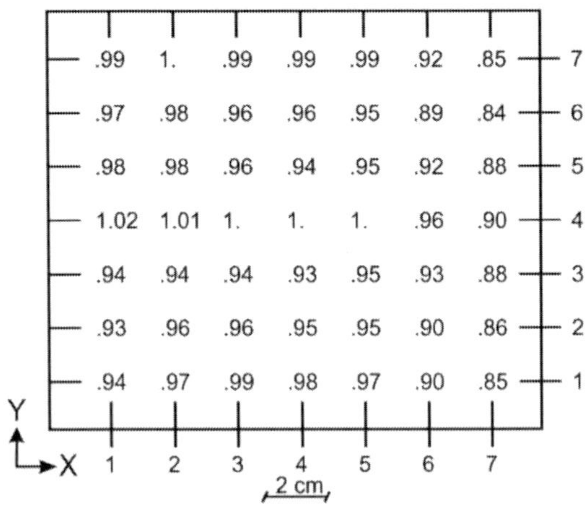

Figure 57. Spread of the amplification coefficient in the MSAC over its area [54].

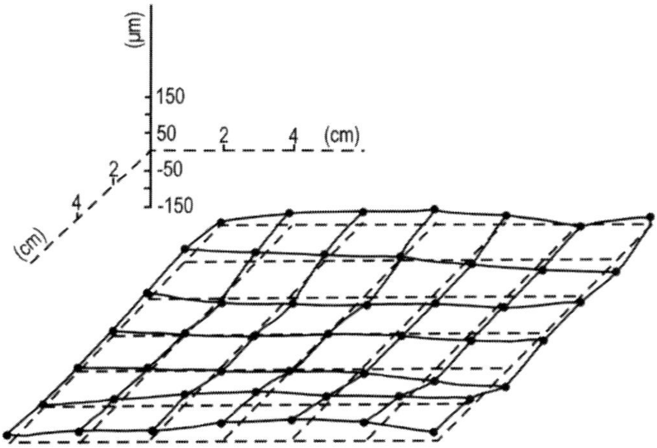

Figure 58. Deviations of the gap width in the MSAC from the ideal plane shown with dashed lines.

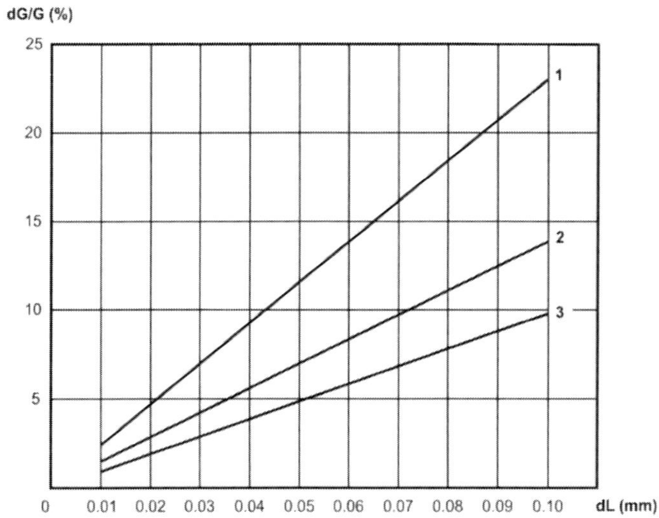

Figure 59. Calculated deviations of the gas amplification coefficient ΔG vs. the spread of the gap with at $G_{pr} = 10^3$ and gap widths of 3, 5 and 7 mm (respectively curves 1,2 and 3).

Figure 59 presents calculated deviations of the gas amplification coefficient ΔG such as versus the deviations of gap width from its nominal value at a fixed value of amplification (Gpr $= 10^3$) and three values of the gap width (3, 5, and 7 mm). From Figure 59, it is apparent that a 100-μm spread of the gap width causes the value of *G* to vary by 10, 14 and 23% for 7-, 5-, and 3-mm-wide gaps, respectively. An increase in gas amplification coefficient *G*pr to a value of 10^5 is followed by an increase in the fluctuations of the gas amplification versus the spread of the gap width. As a result, a 100-μm spread of the gap width causes *G* to vary by 16, 23, and 39% for 7-, 5-, and 3 -mm-wide gaps, respectively. As the preamplification gap width is reduced, more stringent requirements are specified for the accuracy with which the detector components should be produced. This fact is especially important for manufacture of large-area detectors.

3.3. Thermal Neutron Imaging Detectors Combining Novel Composite Foil Convertors and Gaseous Electron Multipliers

The potential of a thermal neutron imaging system based on a composite neutron convertor foil combined with a low-pressure, multistep avalanche chamber demonstrated in the work of Dangendorf et al. [62]. Neutron-induced charged particles from a primary convertor element induce multiple low-energy electrons escaping from a second thin film of high electron-emissive material. Authors investigated the performance of detectors with Gd- and Li-based primary convertors coated with CsI as a secondary electron emitter. It is shown, that the detector can be operated with high stability at a sufficiently high gain to detect all escaping particles, fast time resolution, low occupation time and high count rate A localization resolution of 0.4 mm (FWHM) was obtained. The good imaging performance, free of parallax errors in divergent neutron beams, capability make this instrument an excellent tool for time-resolved neutron scattering experiments and for neutron radiography and tomography.

Idea of original programmers therefore is to replace the single layer convertor with a composite convertor foil; such a foil consists of a traditional neutron convertor coated with a thin film of an efficient secondary electron emitter. Neutron-induced particles from the convertor induce the emission of multiple low-energy secondary electrons, which initiate a well localized avalanche close to the neutron impact location. To obtain the optimal imaging and timing performance it is important for the detector to be sensitive primarily to the slow ionization electrons in the vicinity of the convertor foil. These electrons originate either from gas ionization by the fast particle or from ionizations occurring in the foil, leading to the emission of slow secondary electrons into the gas.

Assuming a mean energy of 50 keV for the primary electron escaping from the Gd into the CsI, authors should expect more than 10 secondary electrons emitted on the average from a few hundred μm thick CsI layer. This number is at least 5 times larger compared to the number of ionization electrons induced in the gas by the 50 keV electron, assuming a 2 mm gap filled with 10 hPa of isobutane. Moreover, due to the statistically distributed ionization processes in the gas and the exponential avalanche growth in the preamplification gap, the effective contribution of these electrons to the final avalanche is even smaller. Thus, the SEE from the CsI is the leading process and authors expect an increase of about an order of magnitude in the pulse-height and detection efficiency close to 100% for fast electrons escaping from the neutron convertor foil.

3.3.1. The Neutron Convertor Foil

The convertor material was deposited on an aluminum disk, 5 mm thick and 75 mm in diameter. A 300 μm thick natural Gd foil was glued onto the Al disk. The Gd surface was coated with CsI layers of various thicknesses, deposited by standard vacuum evaporation. A low-density CsI film of spongy structure was produced by evaporation under 7.5 hPa of Ar. Under this evaporation condition it is known that the deposited layer has a porous structure and its average density is only a few percent of the bulk density of solid CsI. It is expected that this structure, with its large surface-to-volume ratio, leads to an enhanced secondary electron emission. Unfortunately, authors did not have any control over the effective mass of CsI deposited on

Gd, since the thickness monitor did not function under these evaporation conditions.

3.3.2. The Multistep Avalanche Chamber

The preamplification gap is defined by the convertor foil and an 81% transmission grid made of stainless steel. The width of the gap is 2 mm and it is operated in a proportional amplification mode. After preamplification the avalanche of electrons are transferred to the second multiplication region through a 20 mm long drift and diffusion gap, which improves the operation stability at high gains. The second amplification stage, a multiwire proportional counter, consists of a 1-mm spaced wire anode plane, 20 μm in diameter, placed between two 1-mm-spaced, 50 μm in diameter wire cathode planes. The localization of the center of the avalanche corresponding to the neutron-capture location in the convertor foil is obtained by analyzing the induced signals on the cathode wires, applying the well-known delay-line readout technique.

3.3.3. Detector Characteristics

To obtain maximum neutron detection efficiency with Gd-based convertors, the avalanche chamber must be capable of detecting events originating from a small number of slow electrons, often-single ones, produced in the vicinity of the convertor surface. To achieve a count rate plateau, the chamber must therefore operate at a high gas amplification of several times 10^6. To determine the gain capability of the detector the neutron convertor foil was replaced by a polished Al photocathode and irradiated with UV light from an Hg lamp. The avalanches were thus initiated by single primary electrons from the photocathode. Close to 100% of the signals clearly separated from the noise. At high gain of about 4x 10^6, the detector operates in a stable way over long periods, even in the neutron beam.

For all convertors investigated the pulse-height distributions and the total detection efficiencies for neutrons at a wavelength of 0.115 nm were measured at various gas pressures and voltages. Although the detector with the pure Gd convertor was operated at significantly higher gain, the pulse-height spectrum obtained with the composite convertor shows a significant increase in the number of events with medium pulse-heights. This reflects the expected increase in the number of slow electrons deposited in the preamplification gap by SEE from the CsI.

The detector with CsI-coated convertors reaches count-rate saturation at significantly lower voltages, which is important for its long-term operation stability. The efficiency values are about 8%, which is in good agreement with the theoretical value for a thick natural Gd foil under backscattering conditions. The detection efficiencies at 40 hPa seem to exceed the theoretical predictions. However, at this pressure the maximum attainable gain in the preamplification gap is lower by more than a factor of 10 as compared to the value reached at lower pressures. Therefore, the MWPC contribution to the total gain dominates and the detector becomes susceptible to all charges produced somewhere in the gas, making the detector very sensitive to all kinds of background radiation. For example, γ-rays originating from the reactor or from the Gd foil may be Compton-scattered in the detector wall, the emitted electrons may enter the sensitive gas volume and be multiplied and registered. This is an additional argument in favor of the low-pressure operation of avalanche chambers in this application.

Furthermore, it should be noted that for CsI-coated converters the total conversion efficiency is almost independent of the gas pressure, even at the lowest applied pressures where the probability of producing secondary electrons in the first layers of the gas is negligible. This clearly demonstrates the effectiveness of the CsI secondary electron emission.

A spatial resolution of 0.4 mm (FWHM) was obtained.

In the work, authors present a novel neutron-imaging system based on a composite foil convertor combined with a low-pressure gaseous electron amplification device. The system has many attractive features:

- The concept based on comparatively simple and inexpensive technologies.
- The detector is practically not limited in size or geometry.
- A position resolution of better than 0.5 mm (FWHM) can be achieved.
- In divergent neutron beams the position determination is free of parallax errors.
- Due to the thin conversion layer (below 100 μm thickness) the arrival time of thermal neutrons can be determined with an uncertainty of less than 100 ns. This excellent time resolution enables time-of-flight and coincidence measurements.
- An operation at very high count rates (> 10^{-6} s^{-1} mm^{-2}) is possible with adequate read-out and acquisition electronics.

The main drawback of the method is the relatively low detection efficiency compared with ^{3}He or scintillation detectors. The efficiency of the detector was found to be in good agreement with the theoretical predictions of the particle conversion and escape probabilities. It is expected that the application of isotopic pure convertor materials will result in detection efficiencies of 36% and 52% for neutrons with wavelengths of 0.1 nm and 0.2 nm with ^{157}Gd convertors. This can be further improved by a multi-foil system, where n foils are read out by $n + 1$ multistep avalanche chambers, recording electrons emitted from both surfaces of each foil.

3.4. Hybrid Low-Pressure (MSGC) Neutron Detectors

3.4.1. The Detector Principle

A detector scheme with four full size detector segments of 285 **x** 285 mm^2 size developed by Gebauer et al. is shown in Figure 60. Each segment comprises a central composite $^{157}Gd/CsI$ converter foil and either side of the converter in 0.25 mm distance extraction grids embedded in 4.5 mm deep low-pressure gas multiplication gaps. The gaps are closed by two-dimensional position-sensitive MSGC plates, which are fabricated by multi-layer deposition on glass plates [52].

The optimal for thermal neutron detection, whereas for cold neutrons 1–3 μm thick ^{157}Gd converters deliver higher efficiency. The Gd is coated with columnar CsI secondary electron (SE) emitter (SEE) layers delivering a detectable cluster of slow (eV) SEs into the gas volume from a well-defined locus (~ μm) on the converter surface where the fast CE penetrates. The CsI layers have thickness of smaller than 1 μm in order to keep the sensitivity to energetic γ and X-ray background in the beam low. On the other hand, if photons emitted after neutron capture from ^{157}Gd should emit a cluster of SEs triggering an avalanche above threshold, this signal is as valid and as localized as a CE signal due to the thin converter. Between the converter and the extraction grid a high field strength of E = 5–10 kV/cm is applied for enhancing the SE extraction from the columnar CsI surface.

The extraction gap works in addition as a pre-amplification gap with short mean free path for ionization λ_i in the three-stage gas avalanche multiplication mode used for SE amplification. In this exponential gas amplification mode only avalanches starting at the converter surface achieve the full gain, whereas avalanches triggered elsewhere in the gas volume remain below the detection threshold. Therefore, the sensitivity to weakly ionizing X-rays and γ- rays will be significantly reduced.

Due to low-pressure operation at p ~ 20 mbar in the very good, quench gas isobutane very high reduced field strengths E/p and thus exponential gas avalanche multiplication can be achieved over three amplification stages: In the 0.25mm deep pre-amplification gap between converter and extraction grid E/p is as high as 50–500 V/(cm mbar) resulting in $\lambda_i < 53$ μm and thus very good time localization of the SE avalanches. Due to the short λ_i the sensitive detector volume is restricted to the converter surfaces, parallax is avoided and avalanches released from other parts of the detector volume remain below threshold.

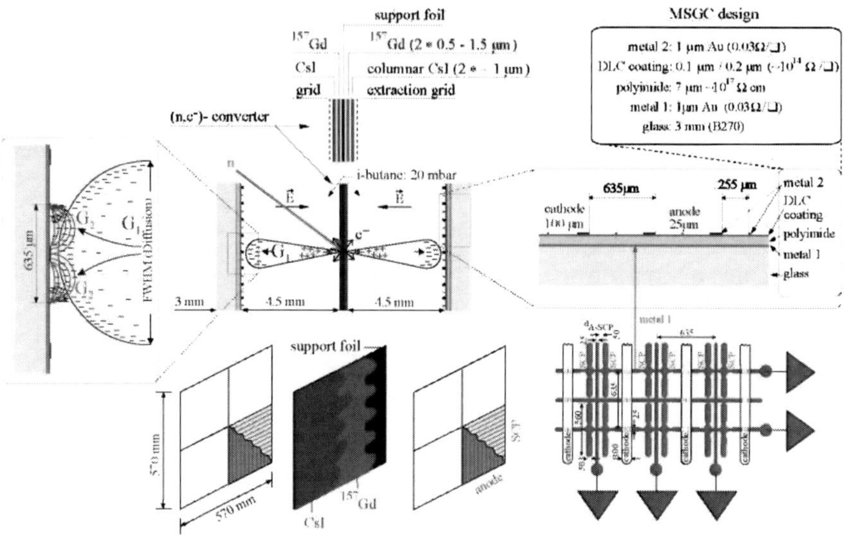

Figure 60. Schematic diagram of a four-segment low-pressure MSGC neutron detector, comprising in each segment (i) a composite $^{157}Gd/CsI$ converter foil with extraction grids; (ii) two adjacent low-pressure (p ~ 20 mbar) gas gaps on either side of the converter with high constant reduced electrical field strengths E/p; (iii) two fourfold subdivided MSGC detector planes, which function as amplification and readout elements. The diagram indicates the gas amplification mode at low gas pressures (left inset), the composite converter foil (top), the micro-strip planes either side of the converter (bottom), some coarse details of the micro-strip plate multi-layer design (right) and the layouts (bottom right) of the micro-strip plane (metal 2), with parallel anode and cathode strips, and of the 'Second Coordinate Pad (SCP)' plane with SCP and return current strips (metal 1, magenta).

In the subsequent constant field region extending from the grid to ~ 0.5 mm distance from the MSGC plate the lower E/p of 100 V/(cm mbar) is still more than sufficient for parallel plate avalanche multiplication. However, only a fraction of 50% of the electrons of the pre-amplification gap can reach this lower-field region by diffusion/scattering through the grid holes. This is enhanced due to the high electron temperature in the high E/p of the pre-amplification gap and due to the only 20μm thick stainless steel extraction grid with double-sided etched holes of 40 μm diameter and 60 μm pitch. This grid reduces further the UV photon feedback from the end of the avalanche to the converter and thus secondary avalanches.

In the high alternating fields between the micro-strip (MS) anodes and cathodes micro-strip amplification sets in with rising E/p when the electron avalanches approach the anodes. Due to strong diffusion extending from the pre-amplification gap over the full trajectory length, the avalanche heads are broadened, and charge is induced on 3–4 anode–cathode pairs although with $\Delta x = 255$ μm a pitch of 635 μm is chosen. Thus, with readout methods determining the center of gravity of the induced charge and with the very good signal-to-noise ratio of the three-stage gas multiplication a position resolution corresponding to a fraction of the MS pitch is achievable. Due to the higher E/p and shorter drift times the count rate capability of low-pressure MSGCs is even higher than that one of normal pressure MSGCs, which exceeds 10^6 cps/mm^2, since most of the positive ions are set free in the gas avalanche close to the anode micro-strips and are drifting over a short distance to the cathode strips.

3.4.2. Converter Fabrication

For fabrication of 1-3 μm thick Gd converters at HMI two half as thick Gd layers were RF sputter-deposited either side on a 6 μm thick Aramid plastic support foil on thin Al barrier layers was used. Great attention was paid to prepare very uniformly stretched Aramid foils because of the high compressive stresses appearing in the thick Gd layers during deposition. In order to avoid thermal deformation of the Aramid foil, for Gd sputtering a two-inch sputter source was operated with low power of 50-100 W. In addition, the rear side of the foil was Peltier-cooled via a polished copper plate. For achieving with the small source a homogeneous layer deposition on full-size plates the substrate must be moved across the source in *x* and *y* directions. This is possible in the new chamber but not in the old setup. Therefore, so far converters for prototype size detectors with 101 **x** 101 mm^2 inner frame opening were made. As depicted in Figure 61, columnar CsI layers of ~ 1 μm thickness and with ~ 4-5 times smaller column base width were made by thermal evaporation in an Ar atmosphere of 10^{-3} mbar base pressure, again on thin Al sub-layers.

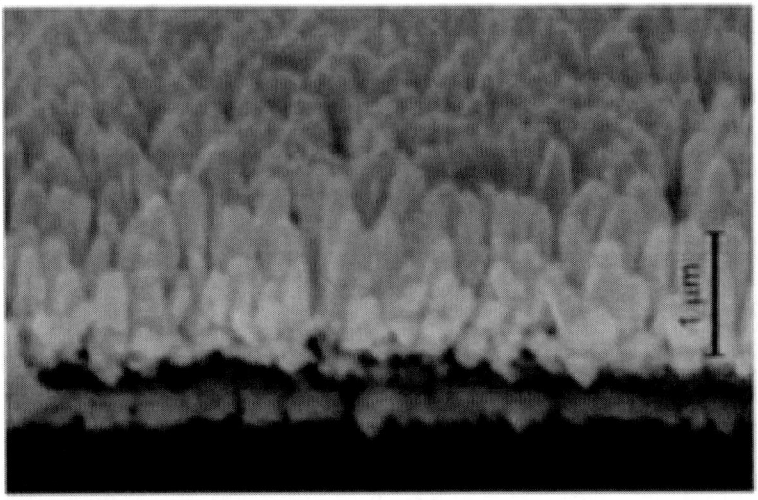

Figure 61. Scanning electron micrograph of a 1mm thick columnar CsI layer deposited as described in the text.

3.5. Resistive Plate Chambers with Gd-Coated Electrodes as Thermal Neutron Detectors

Resistive Plate Chambers (RPCs) basically consist of two bakelite plates kept at a 2 mm distance by a grid of plastic spacers, Abrescia et al. [63]. An appropriate gas mixture is circulated in between and a 4–5 kV/mm electric field is applied. When an ionizing particle crosses the gas gap, subsequent avalanche or streamer processes induce a detectable signal on external readout strips.

Even if this device has been mainly employed to reveal ionizing particles, its possible use to detect neutrons could lead to interesting practical applications like, for instance, the spotting of explosive materials contained in anti-personnel and antitank mines underground. A possible detection technique, accurately studied in the context of the DIAMINE project, uses a ^{252}Cf source placed near the mined ground, which, undergoing fission processes, emits neutrons ranging in energy from about 1 to 4 MeV. Going through the layer of soil overhanging the mine and hitting materials compounding it (mainly H and N), neutrons lose energy in multiple scatterings till termination. The goal is to reveal the backscattered thermal neutrons directing upwards, since an intense enhancement in their number is a precise signature of the presence of the mine. RPCs are very good candidates to fulfill the requirements of an "on-field" application of this technique because they are cheap and mechanically robust.

Neutron detection can only be achieved after interaction with a suitable material, called converter, which has the role of generating ionizing particles. Two gadolinium isotopes, namely Gd and ^{155}Gd, present in absolute, have the largest cross-section to thermal neutrons (of the order of 10^5 barn).

Cross-section for these converters decreasing with *1/v,* where *v* is the velocity of the incoming neutron. This means that Gd can be particularly suited to produce detectors specifically designed to reveal thermal neutrons (and not fast neutrons); this is an advantage, since fast neutrons coming directly from the ^{252}Cf source could constitute a sort of undesirable background.

In the work, natural Gd, which is composed of a mixture of many isotopes, of which ^{157}Gd and ^{155}Gd constitute about 30% of the natural composition, has been chosen. This material was actually used in the form of Gd-oxide (Gd_2O_3), which presents itself as a white inert powder, with granules 1-3 mm in diameter. This is inert, very easy to handle, poses no problem of gas contamination and is very cheap (100$/kg). Gd-oxide powder was put in suspension inside the linseed oil normally coated on the inner surface of the bakelite electrodes, so that the usual method for its polymerisation allows to trap the granules of converter in a solid layer. Since Gd-oxide is not conductive the electric properties of the electrodes are not altered.

To estimate the possible performance of this kind of detector, suitable calculations have been carried out. In particular, a Monte Carlo program has been developed, taking into account the basic interesting processes, i.e., the interaction of the incident thermal neutron, the emission of the secondary electron with the correct energy spectrum, its travel in a Gd-oxide sheet up to its surface and the eventual entering into the detector active volume.

Total detection efficiency rapidly reaches a maximum of around 10–15 μm thickness of about 35%, and after that it decreases slowly. When the Gd-oxide thickness is lower than 10 μm, the probability of an electron stopping inside the Gd sheet is negligible, and the fast rise in efficiency reflects the increasing number of neutron interactions as the Gd thickness increases. When Gd thickness becomes comparable with electron range, one has to take into consideration the fact that neutron flux decreases exponentially inside Gd-oxide; this means that only a surface layer of Gd mostly contributes to the conversion processes.

Once produced, "forward" electrons have to cross more and more Gd-oxide to exit the sheet, as Gd thickness increases, while "backward" electrons have just a, more or less, fixed thickness to cross. This causes two different behaviours for the curves representing the contributions to the total detection efficiency. In this configuration, one can choose to reveal "forward" or "backward" electrons, depending where the detector active volume is located. Revealing "backward" electrons, one has not to worry much about the layer thickness, which may be difficult to control, provided it is greater than 10 μm.

This new technique was applied to build three RPCs, 10 **x** 10 cm^2 in dimension, of which one without Gd_2O_3 and the others coated, on the inner surface of one of the two electrodes, with different Gd_2O_3 concentrations.

Figure 62. Appearance of the detector.

3.6. The Neutron Sensitive Image Plates

Imaging plates (IPs), two dimensional detectors for ionizing radiations such as X-, β-, γ-ray, ultraviolet light, etc. utilizing photo stimulated luminescence (PSL), are used in many fields of applications, such as medical and industrial radiography, autoradiography, X-ray diffraction experiments, and transmission electron microscopy. In each case, the inherent imaging principle is the same explained as follows: after the radiographic image is transferred to an IP, it is scanned point-by-point by a laser beam in an image reader. A series of the PSL emissions corresponding to the scanned pixels is detected by a photomultiplier tube through a high-efficiency light guide to be converted into the electric signals as a function of time. These analog signals are logarithmically amplified and converted into digital signals. By processing these signals through a computer, the computed radiograph can be reconstructed. Moreover, these digital signals can be easily analyzed, stored, and communicated.

Authors have succeeded in experimentally developing an imaging plate neutron detector (IP-ND), Takahashi et al. [64]. The intrinsic part of the IP-ND is composed of a fine mixture of a PSL phosphor, $BaF(Br,I):Eu^{2+}$, and a neutron converter material, Gd_20_3 or LiF. The nuclear reactions of Gd with neutrons produce γ-ray of 0.3, 0.4, 1.2 MeV and other energies and internal conversion electrons with an average energy of about 70 keV, and those of ^6Li produce *a* particles of 2.05 MeV and tritium of 2.74 MeV. These charged particles emitted from the converter can excite the PSL phosphor in the IP-ND, where the ranges of these particles in the phosphor layer are calculated to be less than a few tens of micro-meters, excepting effective penetration length of the γ-ray.

$BaFBr:Eu^{2+}$ is used as a PSL material, and Gd_203 with natural abundance of Gd or ^6LiF is used as a converter material. The average diameters of these powders are 5.0 μm and 3.3 μm or 3.5 μm, respectively. In each IP-ND, 250 μm and 6 μm thick polyethylene terephthalate (PET) films are used as a support and a protection layer, respectively. The size of these IP-NDs is 20 cm X 25 cm or 20 cm X 40 cm. Exposure to neutrons of 2.3 Å wavelength was carried out in JRR-2 of Japan Atomic Energy

Research Institute (JAERI) and the BAS2000, a bio-imaging analyzer system (manufactured by Fuji Photo Film Co., Ltd.), was used for reading the IP-NDs. The pixel size of the BAS2000 can be chosen between 100μ m X 100 μm and 200 μm X 200 μm. The latter was used here.

Image formation process and the image quality factors by means of the IP-NDs, which is similar to the case of X-ray imaging except for the excitation process through the converter. This consideration is essential for neutron radiography as well, in order to design or utilize the system properly.

FCR7000 is an image reader system used in medical diagnostics (manufactured by Fuji Photo Film Co., Ltd.), where a pixel size of 100 μm X 100 μm was used. The result indicated that the ratio of an X-ray photon noise to a light photon noise was about 8:1 at a spatial frequency of 1 cycle/mm. This means that the light photon noise is a minor component of the quantum noises in this system, which is to say that the number of light quanta detected for a single absorbed X-ray photon is sufficiently larger than unity. As for BAS2000, the relative situation among the quantum noises are not different because the laser energy for a unit area and the efficiency of collecting emitted light are almost equal to those of FCR7000.

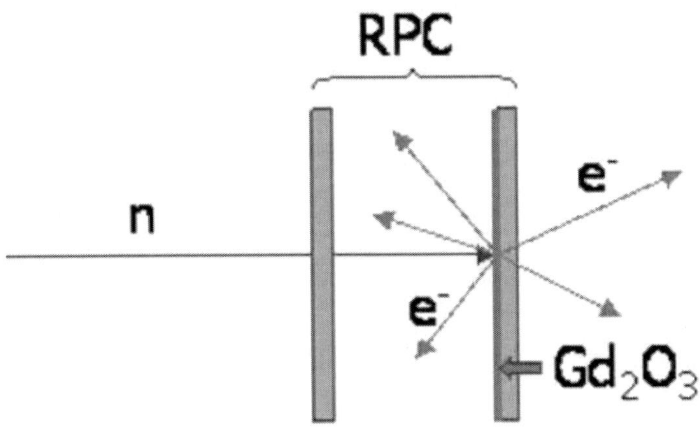

Figure 63. Layout showing the “backward” configuration.

In neutron imaging with the IP-NDs, the energies of secondary particles from Gd and ^{6}Li converters are higher than the average X-ray photon energy of about 40 keV, which corresponds to the exposure tube voltage of 80 kV mentioned above. The number of light quanta detected for a single absorbed-neutron is expected to be sufficiently larger than unity, since the number of light quanta emitted for a single absorbed X-ray photon or neutron can be determined by the energy of the X-ray photon or the secondary particles. Therefore, the light photon noise in the system of BAS2000 and the IP-NDs is estimated to be a minor component of the quantum noises, and the neutron absorption efficiency of the IP-NDs governs the image quality.

The IP-ND has a severe problem in that it is also sensitive to γ-ray since the PSL phosphor is sensitive to X-rays. To solve this problem, an image processing to subtract γ-ray image from the neutron plus γ-ray image was tried. Since Li emits secondary particles of a much higher energy than Gd, the Li-IPs are not so influenced by γ-ray as the Gd-IPs. It is expected that the Li-IPs are more suited to the imaging where there are higher y-ray to neutron ratios than Gd-IPs.

The PSL is maximized at about 50 mol% of Gd to Ba ratio in the Gd-IPs, while it is not saturated up to 90 mol% of 6Li to Ba ratio in the Li-IPs. In case of the same neutron absorption efficiency, both of the Gd-IPs and the Li-IPs show higher PSL under lower converter concentration. It has been shown that most important factor governing the image quality is effective neutron absorption in both of the Gd-IPs and the Li-IPs. For the most part, the Gd-IPs can absorb neutrons more efficiently than Li-IPs. However, the amplification rate in the conversion and excitation process of the Gd-IPs is about 30 times lower. Therefore, the Li-IPs are expected to be more suited to the imaging where there are higher γ-ray to neutron ratios than Gd-IPs.

3.7. Neutron Imaging Detector Using Capillary Phenomena and Liquid Scintillator

Neutron beam and the imaging detectors are useful to investigate the inner feature of human body. So several imaging detectors for neutrons such as an Imaging Plate or scintillating-fiber were developed. In general, three important characteristics are required for neutron imaging detectors. These are fine position resolution, timing resolution, and low background. Though the Neutron Imaging Plate has good position resolution of 25 μm and has high detection efficiency for neutrons, it cannot obtain timing information. In particular, the capillaries filled with liquid scintillator will be good candidate for neutron detector with both fine position resolution and good timing resolution [65]. It is because bundle of capillaries with the diameter of 6 μm can be manufactured at present and the liquid scintillator with Gadolinium (Gd) or Boron (Gd) has high detection efficiency for neutrons. So authors have been developing the neutron detector which consists of liquid scintillator with Gd or B and a capillary plate commercially manufactured by Hamamatsu Photonics Inc. Figure 64 shows the schematic view of the capillary plate and the neutron detector under development.

The capillary plate consists of a bundle of fine glass capillaries with uniform length. The diameter and the length of each capillary are 6-100 μm and 800 μm, respectively. Though the effective area of the typical plate is 20 mm in diameter, it is possible to manufacture the capillary plate with larger diameter. Because the plate is thin, it does not have high detection efficiency for gamma rays which are background for neutron detectors. On the other hand it has good detection efficiency for thermal neutrons in spite of the thinness because Gd or B has high detection efficiency for thermal neutrons. Under the capillary plate absorbing liquid scintillator, optical detector with fine position resolution such as IICCD (Image Intensified CCD camera) or EBCCD (Electron Bombered CCD camera) is attached. As the neutrons are captured by Gd or B, electrons or alpha particles are emitted. They run in the liquid scintillator and the scintillation lights are emitted. If the lights emitted from one capillary by neutron capture can be confined in the capillary and are led to the mounted optical device, the detector can obtain the position resolution of 6 μm.

The key technologies to manufacture such neutron detector are as follows. 1) It is required to absorb the liquid scintillator into each capillary uniformly and to seal it. 2) It is necessary to obtain enough light output by the neutron capture to detect with the optical device. 3) It is required to confine the emitted lights by neutron capture in the capillary and to lead them to the optical device. To realize it, it is necessary to coat the inner surface of the capillaries with light reflector. So we challenged to establish the first technology using capillary phenomena. Then by irradiation with neutrons we investigated the light output from the capillary plate attaching a photomultiplier. Finally, we carried out irradiation experiment of alpha particles attaching the capillary plate with a position sensitive photomultiplier. In this paper, we will report in detail the methods to absorb the liquid scintillator into the capillary plate and results for the basic experiments of neutron and alpha particle irradiation.

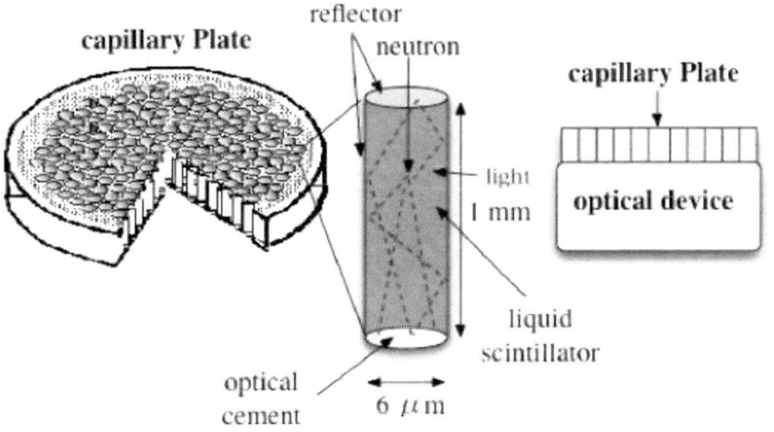

Figure 64. The left figure is the schematic view of the capillary plate. The capillary plate consists of a bundle of fine glass capillaries. The middle figure is the schematic view of one glass capillary. The length and the diameter are about 800 μm and from 6 μm to 100 μm, respectively. Liquid scintillator for neutron capture is absorbed into each capillary. The lights emitted via neutron capture are reflected on the inner surface of each capillary and then are led to an optical device. The right figure is the schematic view of the neutron detector coupled with optical position sensitive device.

Five kinds of liquid were used for these tests. These are water, ethanol, benzene, xylene, and liquid scintillator (BC525) of 2% Gd doped. Authors used two kinds of capillary plates with the diameters of 100 μm and 6 μm, which the side surface inside each capillary is not coated. Though liquid cannot be uniformly absorbed into capillary plates by the method of Type (A) for any combination, we succeeded in absorbing any liquid into capillary plates by the method of Type (B).

Authors have been developing the neutron detector with fine position resolution by absorbing liquid scintillator with Gd or B and to the capillary plate by capillary phenomena. They have established the methods to uniformly absorb the liquid scintillator to each capillary and to seal both sides. Injecting neutrons to the capillary plate attached on the PMT, the basic characteristics have been investigated. From the results, it was confirmed that the capillary plate with the liquid scintillator of BC525 can operate as the neutron detector though the light output is not yet enough. Moreover, injecting alpha particles to the capillary plate attached on the PSPMT, prototype position sensitive detector has been tested. However, the position resolution is not good yet because the lights emitted from the capillary are not confined in the capillary. In near future, will try to manufacture the capillary plate with the liquid scintillator of Boron doped and will establish the method to plate the inner surface of capillaries with light metal for the light reflector.

3.8. Position Sensitive Detection of Thermal Neutrons with Solid State Detectors (Gd Si Planar Detectors)

The central concept of these new detectors is to combine position sensitive Si planar sensors with external Gd converter foils, Bruckner et al. [14]. The thermal neutrons are absorbed in the Gd foil and the resulting conversion electrons detected in the silicon devices. Figure 65 shows a schematic drawing of a Gd-Si neutron detector consisting of one Gd converter foil and two Si detectors. A position sensitive Si planar detector consists of an array of many p^+n-diodes on one single substrate. Each p^+n-junction works as a detector for ionizing particles. The position of the detected particle is given by the individual address of the diode that collects the charge. Each diode is equipped with an individual amplifier chain to ensure real time readout. Typical wafers have a diameter of 4-6 in. and a thickness of about 300 μm. The single structures can be as small as 10 μm and are limited in their extent only by the size of the wafer. A position resolution of less than 10 μm has been achieved with these devices, as well as the construction of large detectors with areas up to m^2, for several high-energy physics experiments.

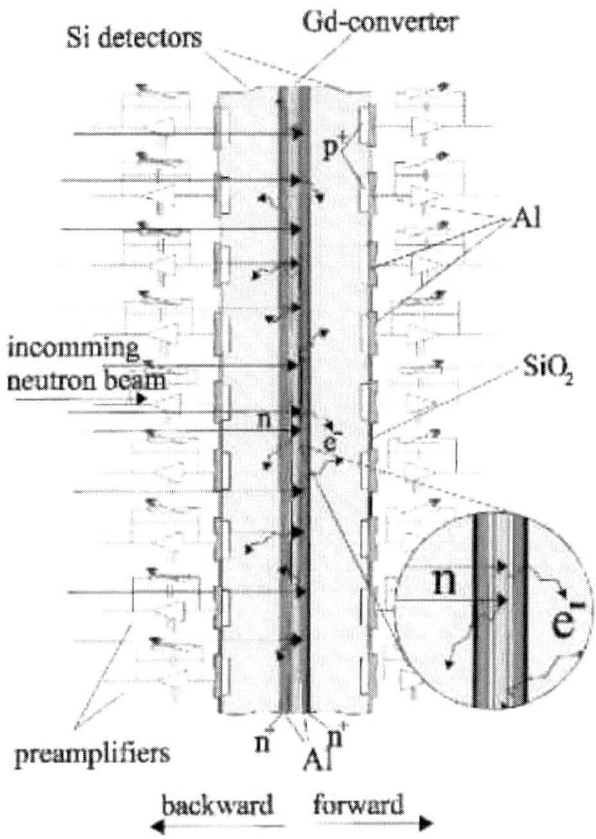

Figure 65. Schematic of a 4% Gd-Si-neutron detector consisting of one converter foil and two Si-detectors with corresponding electronics.

Considering only one converter layer (no sandwich detectors), the (n, γ) reaction in Gd, which produces also a reasonable number of conversion electrons, is the only suitable one for high detector efficiency. The proton, α-particle and triton from the ^{6}Li and ^{10}B reactions have a very short range. Therefore, these converters can only be used in situations, which require low detection efficiency, e.g. beam monitors, or for the detection of cold or ultra-cold neutrons. One can also think of doping the Si detectors with Li, but the usual doping concentrations of about $10^{16}/cm^3$ are much too low for efficient absorption of thermal neutrons.

The (n,γ) reaction in Gd produces a quite complex conversion electron spectra with significant energy lines between 29 and about 200 keV. Due to the high absorption, cross section of Gd the converter can be made very thin, so that the range of the conversion electrons is sufficient to reach the Si-detector. The band-gap in Si of 1.11 eV results in an average energy of about 3.6 keV for the electron hole creation. Therefore, the device must be capable to measure a charge of less than 5000 e^- to detect the low-energy conversion electrons, which have already lost part of their energy when escaping the converter foil.

The detectors are made of high resistivity silicon in planar technology and operated in fully depleted mode. For a typical detector less than 100 V applied to the p^+ n-junction ensures, that the whole detector works as one large depletion region. This mode decreases the detector noise, because it reduces the capacity between strip and backplane, and the thermally generated leakage currents. In addition, it guarantees that the entrance window on the rear side is as thin as the one on the front side and that there are no dead sections between the strips. While CCDs are made of only partly sensitive shift registers and contain inefficient channel stops to separate the columns, Si planar detectors are continuously efficient devices. Another important distinction between these detectors and CCDs concerns their readout. CCDs are integrating devices with charge transfer that means serial readout. Therefore, high-resolution time information is very difficult to obtain. Contrary Si detectors constructed to measure single events with parallel readout electronics. Provided the lower level discriminator is set sufficiently high, the energy distribution of the conversion electrons does not affect the detector response, as in the case when the spectra would be detected by an integrating device such as film or CCD.

The theoretical time resolution of the Gd Si detector given by the charge collection time in Si, the transit time of the thermal neutrons through the converter foil and the lifetime of the excited Gd atoms after the neutron capture. None of these intervals is longer than some nanoseconds, which allows the construction of a very fast detector.

For parallel readout, it is necessary to equip each detector diode with the full amplification and readout electronics (more sophisticated readout techniques can reduce the number of channels). For many channels and/or a high position resolution, this can only do by the use of very large-scale integrated electronics. Although the technique used for the detector fabrication from a general point of view similar to the one used for electronic circuits, it differs in some essential aspects, which prevent so far the construction of detector and electronics on the same wafer.

3.8.1. Characteristics of Detector

For every stopped neutron the probability, that an electron will be emitted is given by the conversion factors of the corresponding (n, γ) reaction. As the decay pattern of Gd is quite complex some simplifications have to be made for the calculations of the detector efficiency. Taking the values of the dominant isotopes ^{155}Gd and ^{157}Gd from the nuclear data sheets, we used only the lowest energy levels which have the highest conversion factors, E_y (keV) = 88.97, 199.21, 296.53, 263.58 for 156Gd and E_y (keV) = 79.51, 181.93, 277.54, 255.66 for 158Gd. The energy distribution of the conversion electrons was calculated, by subtracting the binding energies of the K, L and M shells. Although each conversion electron is accompanied by several γ-rays as well as X-rays and Auger electrons, which follow the ionization of the atom, only the conversion electrons, and the Auger electrons with an energy of more than 20 keV have been taken into account. For all electrons, which enter the Si detector with a remaining energy of at least 15 keV, the efficiency of the Si detector can be assumed to be 100%.

The efficiency, defined as the number of registered neutrons divided by the number of incident neutrons, was measured in backward direction for six different wavelengths in comparison to a calibrated ^{3}He counter. The results for natural Gd and an enriched (90.5%) ^{157}Gd converter are shown in Figure 11. For thermal neutrons, authors can conclude an efficiency of more than 35% for the enriched ^{157}Gd converter and about half of this value for natural Gd. One can clearly see the increase towards higher wavelengths.

The measured values are constantly higher than the calculated ones. That this is caused by X-rays and γ-rays following the neutron capture, which are partly detected. The X-ray lines is more dominant for natural Gd. Conversion electrons, γ-rays and X-rays are emitted simultaneously compared to the time resolution of the amplifier of about 1 μs, therefore, one captured neutron can cause only one signal in the detector. Nevertheless, the electromagnetic radiation can increase the overall efficiency in the case when the electron not detected. Otherwise, the detection of γ-rays and X-rays causes only a spread of the energy spectra.

With the first measurements about three years ago, authors could demonstrate that a one-dimensional neutron detector can be constructed from a single-sided strip detector. The position resolution achieved with this early system is compared with the results of a pad detector and the latest measurements with a double-sided strip detector.

As for the previous setup the position resolution was determined by measuring the detector response to an edge. A Lorentzian edge spread function was fitted to the data near the edge, which gives a position resolution of FWHM= 210 μm.

The measurements confirmed that the distance between Gd converter foil and detector is the critical parameter for the position resolution. To take advantage of the extremely high resolution of the strip detector, the Gd converter has to be placed onto the detector surface or deposited directly on the Si wafer.

3.9. New High-Resolution Gadolinium-GEM Neutron Detectors

The detector discussed in Pfelffer et al. [15] is shown in Figure 66. The detector under test was a standard 10 x 10 cm^2 Triple-GEM [72] detector flushed with Ar/CO2 70/30 mixture at 10 l/h at room temperature and atmospheric pressure. It consisted of three GEM foils [73] stacked at a distance of 2mm from each other and powered with a resistor chain. The detector was operated at an effective gain of about 5000. The conversion volume or drift space was 10mm long, and as cathode served a 250 μm thick sheet of natural gadolinium. The high voltage to the cathode was provided with a second power supply. A drift field of 700 V/cm was chosen to avoid electron attachment to electronegative impurities and the subsequent loss of primary ionization electrons, while keeping the drift velocity smaller than 2.0 cm/μs [74] at atmospheric pressure. The readout board was a Cartesian x/y strip readout [72, 75] with 256 strips and 400 μm pitch in x and y direction. Four APV-25 hybrid chips per readout were used to pre-amplify the signals. The APVs were operated in multi-mode, and the signal from the bottom of the third GEM foil served as external trigger signal. In APV multi-mode, three consecutive pipeline columns are reserved each time the chip is triggered. Each column is then read out separately in peak readout mode. The waveforms were digitized with the Scalable Readout System (SRS) and acquired with the ALICE DAQ system DATE and the Automatic MOnitoRing Environment (AMORE) software.

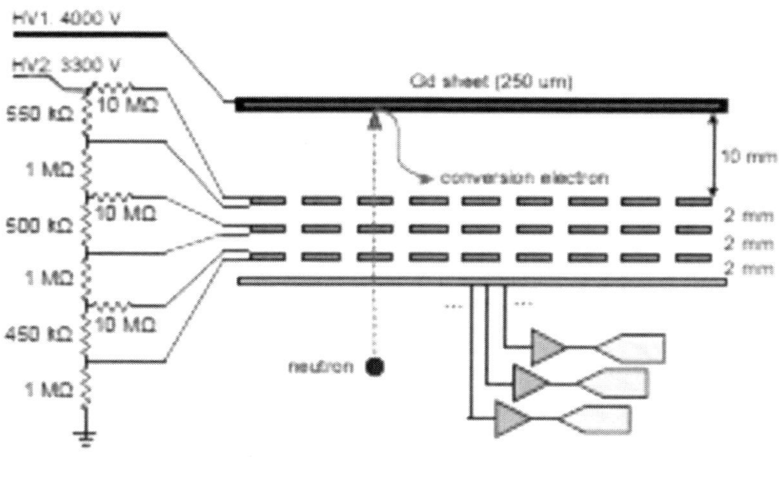

Figure 66. Schematic representation of the detector.

Figure 67. Measurement setup at R2D2 beam line with collimation slits and Triple-GEM detector in backwards configuration.

For the detector tests at JEEP II, the monochromator of the R2D2 beam line was optimized to deliver mono-energetic neutrons of 2 Å. A multi-slit Gd_2O_3 Soller collimator that limits the angular divergence of the beam to 30' is inserted into the R2D2 shutter. Figure 67 shows the beam line behind the shutter. It was equipped with a beam monitor, two sets of movable collimating slits made of borated Al blades, and a calibrated Reuter-Stokes 1 inch ^3He tube (model RS-P4-0810-220). The tube was operated at a ^3He pressure of 10 atm, and a voltage of 1400V. The walls of the tube are made of stainless steel and have a thickness of 0.508 mm. At a wavelength of 2.5 Å, the efficiency of the tube is (96.3 ± 0:3)%. From this efficiency, the efficiency for neutrons with a wavelength of 2 Å was extrapolated to (95 ± 0.5)%.

Detector shown efficiency 11.8% for neutron for 2 Å and good spatial resolution.

3.10. USE GADOLINIUM AS ADDITIVES IN SCINTILLATORS

A large review of the use of Gadolinium as additives in plastic and liquid scintillators is given in the paper by Dumazert and Coulon [76].

Compared to coating techniques, the interest of Gd loading lies within the maximization of the (n,γ) interaction rate, as well as the amplitude of the signature, for a given mass of the incorporated element. Inorganic Gd-scintillators, such as Ce^{3+}–doped GdI_3 crystals, have been described to have excellent light yield (expressed in photon per megaelectronvolt, ph/MeV) and neutron response under exposition to a moderated Cf-252 source. Out of scalability, shape versatility and manufacturing cost considerations, however, the incorporation of Gd in plastic scintillators is more common, either by homogeneous or heterogeneous loading.

Plastic loading with heavy elements, such as Gd, is mainly limited by the lack of solubility of such dopants into the hydrocarbon-based solvent. This drawback may lead to the degradation of the material's optical properties, notably through turbidity. For a homogeneous loading, the insertion of heavy elements in the form of organometallic compounds, with at least one covalent bond between the metal and the organic group (C, H, O), is required. Reference work by Czirr presents the use of a Gd-benzoylacetonate ($Gd(Hba_3)$) complex for the loading of a commercial, polyvinyltoluene (PVT) NE120 scintillator (Nuclear Enterprises). Research was then undertaken to overcome the maximum 0.5 wt% load introduced the use of a hexamethylphosphotriamide (HMPT) co-dopant, which has enabled 3 wt% metal loading of polymethylmethacrylate (PMMA) under the form of Gd nitrate. Expected registration efficiencies in both cases were estimated between 12% and 15% for neutrons below 500 meV. Preference was given to organometallic compounds with aromatic units, such as Gd-phenylcarboxylates in polystyrene (PS) matrices: gadolinium-5-phenylpentanate (or) with solubilizer triphenylphosphine oxide (TPPO) was inserted up to 50 wt% complex.

Reference work by Bell et al. [77] demonstrated the relevance of Gd-loaded PS for neutron detection. A PVT matrix was loaded up to 12 wt% with a tributylphosphate (TBP) complex of Gd nitrate, with satisfactory material (colorless samples) and optical (less than 50% light output degradation) features. Irradiation of 5 cm^3 samples by a 100 $n*cm^{-2}*s^{-1}$ thermal neutron flux ($6\cdot10^5$ $n*s^{-1}$ Am-Li source behind a W shield) resulted in a response spectrum with discriminable peaks centered at 38 keV and 80 keV. The peaks are identifiable with a convolution of the main photon and electron signatures below 100 keV.

The complex was initially introduced in the loading of high-scale (10 m^3) liquid scintillators for neutron background suppression in power plants, and electron antineutrino detection experiments [78].

In the field of material characterization by photofission, Beil et al. [79] designed a very-large-scale neutron counter based on a 0.5 wt% Gd-loaded commercial, xylene NE323 (Nuclear Enterprises) liquid scintillator. Monte Carlo calculations have shown 60% to 85% registration efficiency, in accordance to a tradeoff between neutron capture and gamma background.

Xylene-based scintillators induce safety hazards because of potential toxic matter leakage and relatively low-temperature flash point (38°C for NE323). As a consequence, considerable effort went over the last decade into less flammable and less toxic chemical formulations, to work large-scale liquid scintillators into practical homogenous detection media. This research was notably carried out in the framework of reactor antineutrino experiments. In this context, antineutrinos first interact with hydrogen, present in large amount inside the target, according to the inverse beta decay reaction equation:

$$ {}^{0}_{0}\overline{\nu}_e + {}^{1}_{1}p^{+} \rightarrow {}^{1}_{0}n + {}^{0}_{1}\beta^{+} $$

An antineutrino is thus detected through the annihilation of the positron and the delayed capture of the neutron, with the temporal correlation of both signals increasing the ratio between antineutrino signature and random background.

Loading Gd in the liquid scintillator devoted to neutron counting has been found to significantly reduce the mean time gap between antineutrino reaction and neutron capture, therefore increasing the correlation frequency rate and lowering the false detection background. The preparation, developed in the frame of the Double Chooz reactor antineutrino experiment, was based on a linear dodecane ($C_{12}H_{26}$, 80%) solvent mixed with ortho-phenylxylylethane (o-PXE, $C_{16}H_{18}$, 20%), and showed high attenuation length at 430 nm (7.8 ± 0.5 m), a critical feature for the collection of the scintillation signal in a large-scale radiation sensor. Background is kept below 1% of the expected neutron signal by applying a frequency high-pass filter, with a threshold at 10 Hz, on the count rate.

As an alternative to Gd-loaded scintillators, Dazeley et al. [80] investigated another high-scale liquid detector, in the form of a 250-L, 0.2-wt% $GdCl_3$-doped water Cherenkov tank. A set of 8 UV-sensitive photomultiplier tubes allows for a resulting 10-% collection efficiency of Cherenkov photons. The signature above 3 MeV equivalent is filtered in the event-frequency space on the basis of an expected inter-neutron capture period of about 28 μs. This spectral treatment allows for a rejection of high-energy fission gamma rays (such as the prompt gamma rays from Cf-252 spontaneous fissions, used in the experimental study) and background. Only spallation neutrons from the natural muon shower and undergoing capture form a background that cannot be discriminated in the frequency space from the Cf-252 source signature. The concept of this very-large-scale neutron was transposed to the field of SNM passive detection by means of a GEANT4-based optimization study [81]. The aim of the study is to design 1-tonne water Cherenkov modules covering a large solid angle. The counting algorithm aims at detecting time correlated neutrons, i.e., multiple neutrons from a single fission (or even subsequent neutrons from fission chains in a weakly multiplying source), to reduce the rate of false alarms. The complete design of a module includes nine PMTs, a segmentation between a Gd-loaded region (1/5) and an unloaded region (4/5), and a top muon veto. A high-energy threshold (140) photoelectrons, corresponding to 2.6 MeV gamma ray total absorption) accounts for gamma background without pile-up, and the detector segmentation serves the rejection of spallation neutron detection. With a frequency threshold at 130 Hz over triggering events, a 4-module water Cherenkov system with a 0.6 wt% $GdCl_3$ load is expected to have a 25% registration efficiency and a detection limit of 75 g of plutonium within 20 s.

CONCLUSION

Theoretical bases of conversion electrons formation at reaction of radiating capture of neutrons are considered. Based on Bloch theory brake ability of electrons is viewed. Model calculations of efficiency of registration of thermal neutrons by the foil converters made from natural gadolinium and its 157 isotope carried out. Processes of neutron absorption in the material of a converter and the probability of secondary electron escapes were examined. Calculation made for converters with the various thicknesses. We have chosen the most optimal converter thicknesses, both from natural gadolinium and from its 157 isotope. While using converters from natural gadolinium it is possible to obtain total efficiency of 10, 21, 26, 30%, correspondingly for the neutrons with wavelengths of 1, 1.8, 3 and 4 Å with converter thickness 24, 7, 5, 4 microns. For the converters with 157 isotopes it is possible to reach total efficiency of registration up to 27, 45, 49, 52% for the neutrons with the wavelengths 1, 1.8, 3 and 4 Å with the thickness of converter 5, 3, 2, 2 microns.

In earlier calculations, we took into account only electrons with energy higher than 29 keV. We are not taken into account the low-energy Auger electrons, the electrons from L-subshell with the energy 4.84 keV and M-subsell with the energy 0.97 keV. These electrons have rather small free path length in gadolinium; these are 0.3 microns (4.84 keV) and 0.04 microns (0.97 keV). They bring the small contribution to the general efficiency at use of converters made from natural gadolinium as the length of free path of neutrons in natural gadolinium makes tens micron. At the same time, their contribution becomes essential at use of converters made from 157 isotope of gadolinium as the free path length of neutrons in them does not exceed 2-3 microns and this length becomes comparable with length of electron paths.

Calculations of the efficiency of registration of thermal neutrons by foils made from natural gadolinium and its 157 isotope carried out. In calculations, the low-energy electrons were taken into account. The received results are in a good agreement with the experimental data.

We have estimated the contribution of X-ray and low-energy gamma-quanta absorbed directly in a material of the converter and resulting in occurrence of secondary electrons. At practical calculations, it is also necessary to take into account the absorption of quanta in materials of detectors. At such calculation, total efficiency can appear higher, as the part of electrons can formed from quanta in a material of the detector. In our calculations, we take into account only an output of electrons from a material of the converter. In case of the account of the contribution of electrons formed by X-ray quanta the efficiency increased a little, but their contribution is insignificant. The occurred increase is no more, than by 1%. It is rather possible, that X-ray quanta can affect essentially to the total efficiency at the account of their absorption in a material of detectors. Therefore, gas detectors with Argon filling can increase the efficiency essentially. If the conversion of quanta will occur in a working body of the detector, therefore the full gathering of formed secondary electrons will take place. They can render even greater influences in the case of use of semi-conductor detectors.

Calculations of complex converters representing a set of thin gadolinium foils located on the both sides of supporting kapton foils carried out. These calculations are of interest from the point of view of calculation of GEM structures. They have the similar geometry, but only contain apertures in 300 microns diameter and with 300 microns step. Thickness the kapton foil considerably influences on the efficiency of neutron registration. With the increase of thickness of a film the probability of the secondary electron escape decreases. These kinds of converters could give a good result at application of natural gadolinium.

The new solid-state converter of thermal neutrons offered. The converter will consist of a set of thin gadolinium foils located one over other in a gas volume. Foils there will drilled with the fine step (2 mm) with diameter of apertures 1 mm. These foils will have an optical transparency of 40%; correspondingly, gadolinium will fill 60% of a surface. Secondary electrons will emit for all sides, and in an electric field, they will be soaked up in apertures and further drift in the direction of the detector. As a detector there a various gas detectors, such as the multi-wire proportional chamber, multi-step avalanche and multi-strip detectors, and so on, can used. Calculations for this complex converter representing a set of thin gadolinium foils located one over other in a gas volume made. Similar converters can give good result at application of foils from natural gadolinium. Technologically, foil thickness should be more than 5 microns. Thinner films are difficult to drill.

In book described the position - sensitive detector of thermal neutron radiation. This detectors use the solid-state converters such as normal-pressure multistep avalanche chambers (MSAC), low-pressure multistep avalanche chambers (LMSAC), microchannel plates, thin layer scintillators. New types of detectors are offered and developed such as hybrid low-pressure micro-strip gas chamber (MSGC), position sensitive silicon detectors (Gd Si PD), resistive plate chambers (RPCs), GEM structure detectors, imaging plate neutron detectors (IP-NDs), plastic and liquid scintillator, liquid scintillator into a capillary plate and so on.

Basic characteristics of MSAC described. The factor of preliminary amplification of MSAC can reach value 10^5. However, for reception of a stable mode without sparking it is necessary to choose it $<10^4$. The full factor of gas amplification of detector G makes $10^6 \div 10^7$. At $G > 10^6$ registration of individual photoelectrons is possible. Extent of a plateau counting rate characteristic at registration gamma-quantum's with energy 6 keV makes 800 V. MSAC have extent of a plateau counting rate characteristic 300 V for registration β-particles. Efficiency of registration of relativistic particles close to 100%. The energy resolution at registration of gamma-quantums from a source ^{55}Fe is equal 18%. The time resolution at use of mixes based on argon $20 \div 30$ nanoseconds (FWHM) and 15 nanoseconds (FWHM) at use of neon mixes. The own spatial resolution of the detector equal of 260 microns (FWHM) and 100 microns (σ). Spatial resolution MSAC at registration gamma - quantum's with energy 6 keV about 400 microns (FWHM). The opportunity of improvement of spatial resolution of MSAC in case of registration non-collimated charged radiation by peak selection of events is shown. Thus on analysis of amplitude of signals it is possible to judge above the angle of an input of particles in volume of the chamber. Application of this methods - "electronic collimation" - allows to improve approximately twice the spatial resolution. The method can be applied at registration of conversion electrons, escaping out from converters at registration neutron and gamma - radiations.

Dependence of characteristics of MSAC on concentration of molecular additives investigated. So change of concentration of additives on ± 0,5% causes change G in ± 10 times. Extent of a plateau strongly depends on change of concentration of the additive as counting characteristics and time resolution MSAC. Ways of stabilization of operating mode MSAC, which at use allow reaching long-term stability of work of the detector, are developed.

ACKNOWLEDGMENTS

Mathematical modeling of this work has accomplished under the Project T-1157 of the International Science and Technology Center. I would like to thank Professor H. Rauch and Doctor V. Dangendorf for interest to our work.

REFERENCES

[1] Convert, P; Forsyth, JB. (eds.), *Position-sensitive detection of thermal neutrons*, Academic Press, London, 1983.

[2] Meardon, BH; Salter, DC. *A Survey of Position Sensitive Detectors and Multi-Counter Arrays with Particular Reference to Thermal Neutron Scattering*, 1972, *RHEL-R* -262, 91.

[3] Crane, TW; Baker, MP. Chapter 13, "Neutron Detectors", in *Passive Nondestructive Assay of Nuclear Materials*, edited by TD; Reilly, N; Ensslin, HA. Smith, US Nuclear Regulatory Commission NUREG/CR-5550, March 1991.

[4] Rauch, H; Grass, F; Feigl, B. "Ein neuartiger fur langsame neutronen" ["A novel for slow neutrons"], *Nuclear Instruments and Methods in Physics Research*, 1967, 46, 150-153.

[5] Feigl, B; Rauch, H. "Der Gd-neutronenzahler" ["The Gd-neutron counterpart"], *Nuclear Instruments and Methods in Physics Research*, 1968, 61, 349-356.

[6] Jeavons, AP; Ford, NL; Lindberg, B; Sachot, R. "A new position-sensitive detector for thermal and epithermal neutrons", *IEEE Transaction on Nuclear. Science*, 1978, Vol.NS-25, No 1, 553-556.

[7] Charpak, G; Sauli, F. "The multistep avalanche chamber: a new high-rate, high-accuracy gaseous detector", *Phys. Letters*, 1978, v. 78B, № 4, 523.

[8] Melchart, G; Charpak, G; Sauli, F. "The multistep avalanche chamber as a detector for thermal neutrons", *Nuclear Instruments and Methods in Physics Research*, 1981, 186, 613.

[9] Abdushukurov, DA; Djuraev, AA; Evteeva, SS. et al. "Model Calculation of Gadolinium- Based Converters of Thermal Neutrons". *Nuclear Instruments and Methods in Physics Research*, 1994, v. B 84, 400.

[10] Abdushukurov, DA. "Gadolinium Foils as Converters of Thermal Neutrons in Detectors of Nuclear Radiation", *Series: Physics Research and Technology,* ISBN: 978-1-61728-676-6, New-York, Nova Science Publishers Inc., 2010, pp. 1-124.

[11] Abdushukurov, DA. "Application of the Gadolinium Foils as Converters of Thermal Neutrons in Detectors of Nuclear Radiation". In: Caden C Tompson editor, "Gadolinium: Compounds, Production and Application", Chapter 2, *Series: Chemical Engineering Methods and Technology*, ISBN: 978-1-61668-991-9, New-York, Nova Science Publishers Inc., 2010, pp. 53-141.

[12] Abdushukurov, DA; Abduvokhidov, MA; Bondarenko, DV. et al. "Modeling the registration efficiency of thermal neutrons by gadolinium foils", *J. of Instrumentation, 2007*, JINST, 2, PO4001, 1-13, Archives of Los Alamos National Laboratory USA, 2007, 1-19. http://arxiv.org/ftp/physics/papers/0611/0611225.pdf.

[13] Abdushukurov, DA; Bondarenko, DV; Chistyakov, DY. "Contribution of nano-scale effects to the total efficiency of converters of thermal neutrons on the basis of gadolinium foils". *Archives of Los Alamos National Laboratory USA*, 2008, 1-9. http://xxx.lanl.gov/ftp/arxiv/papers/0802/0802.0401.pdf.

[14] Bruckner, G. et al. "Position sensitive detection of thermal neutrons with solid state detectors (Gd, Si planar detectors)", *Nucl. Instrum. Meth.* A 424 (1999) 183.

[15] Pfeiffer, D; Resnati, F; Birch J. et al. "First measurements with new high-resolution gadolinium-GEM neutron detectors", *JINST,* 11, P05011, 2016, 1-16, doi:10.1088/1748-0221/11/05/P05011.

[16] Harms, AA; and McCormack, G. "Isotopic conversion in Gadolinium-exposure neutron imaging", *Nucl. Instrum. Meth.* 118 (1974) 583.

[17] Bäcklin, A. et al. "Levels in 156 Gd studied in the (n; γ) reaction", *Nucl. Phys.* A 380 (1982) 189.

[18] Greenwood, RC. et al. "Collective and two-quasiparticle states in 158 Gd observed through study of radioactive neutron capture in 157 Gd", *Nucl. Phys.* A 304 (1978) 327.

[19] Hargrove, C.K; Blevis, I; Paterson, D; and Earle, ED. "Use of Gd in gas counters as neutron detectors in SNO", *Nucl. Instrum. Meth.* A 357 (1995) 157.

[20] Abdushukurov, DA. "Mathematical Modeling of the Efficiency Gadolinium Based Neutron Converters", *Appl. Math.* 4 (2013) 27.

[21] Gusev, NG; Dmitriev, PP. "Quantum radiation of radioactive nuclear", Moscow, *Energoatomizdat*, 1977.

[22] Kozlov, VF. "Manual for Radiation Safety", Moscow, *Energoatomizdat*, 1987.

[23] "*Table of Radioactive Isotopes*", Radiochemistry society, http://www.radiochemistry.org/periodictable/frames/isotopes_lbnl/

[24] Under edition Kikoin, I.K., "The table of physical parameters", Moscow, *Energoatomizdat*, 1976.

[25] "Evaluated Nuclear Data File (ENDF)", *IAEA-NDS*, http://www.nndc.bnl.gov/exfor3/endf00.htm.

[26] "Thermal Neutron Capture Gammas by Target", *NDS*, IAEA, http://www-nds.iaea.org/oldwallet/tnc/ngtblcontentbyn.shtml.

[27] Hahn, O; Meitner, L. *Z. Phys.*, 1924, 29, 169.

[28] Pauli, HC; Alder, K; Steffen, RM. "The Theory of Internal Conversion." In *The Electromagnetic Interaction in Nuclear Spectroscopy*, edited by D. Hamilton, North-Holland, Amsterdam, 1975, Chap.10.

[29] Rosel, F; Fries, HM; Alder, K; Pauli, HC. "Internal Conversion Coefficients for all Atomic Shells", *Atomic Data and Nuclear Data Tables*, 1978, vol. 21, 91.

[30] Lee, MA. "Nuclear data sheets for A=158", *Nuclear Data Sheets*, 1989, 56, 158.

[31] Bricc 2.0a. "Band-Raman International Conversion Coefficients", *BNL,* http://www.nndc.bnl.gov/bricc/

[32] Lederer, CM; Shirley, VS. "*Table of Isotopes*", 7th Edition, Wiley, New York, 1978.

[33] Chen, MH; Crasemann, B; Mark, H. *Atomic Data and Nucl. Data Tables*, 1979, vol. 24, 13.

[34] Krause, MO. *J. Phys. Chem.*, Ref. Data, 1979, vol.8, 307.

[35] Dillman, LT. "EDISTR - A Computer Program to Obtain a Nuclear Decay Data Base for Radiation Dosimetry", *Oak Ridge National Lab. Report ORNL*/TM-6689, 1980.

[36] Larkins, FB. *Atomic Data and Nucl. Data Tables*, 1977, vol. 20, 313.

[37] Bethe, HA; Ashkin, J. "Passage of radiations through matter", *Experimental Nuclear Physics*, Wiley, New York., 1953, vol.1, 166.

[38] Sternheimer, RM. "The density effect for the ionization loss in various materials", *Phys.Rev.*, 88, 851, 1952.

[39] Uehling, EA. "Penetration of heavy charged particles in matter", *Annual Rev. Nucl. Sci.*, 1954, No 4, 315.

[40] Conen, ER; Taylor, BN. "The 1973 least-squares adjustment of the fundamental constants", *J. Phys. Chem. Ref. Data*, 1973, No 2, 663.

[41] Moller, C. "Zur Theorie des Durchgangs schneller Elektronen durch Materiie" ["Theory of passing fast electrons through matter"], *Ann. Phys.*, 1932, vol. 14, 568.

[42] Rohrlich, F; Carlson, BC. "Positron-electron differences in energy loss and multiple scattering", *Phys. Rev.*, 1953, vol. 93, 38.

[43] Inokuti, M. "Inelastic collisions of fast charged particles with atoms and molecules- the Bethe theory revisited", *Rev. Mod. Phys.*, 1971, vol.43, 297.

[44] Bichsel, H. "Passage of charged particles through matter", in *American Institute of Physics Handbook*, 3rd Edition, D.E.Gray ed., New-York, 1972, 8.

[45] Walske, MC. "Stopping power of L-electrons", *Phys. Rev.*, 1956, vol. 101, 940.

[46] International Commision on Radiation Units and Measurements, *"Stopping Powers for Electrons and Positrons",* 1984, ICRU Re 37.

[47] Murin, AN. *Physics Principal of Radiochemistry*, Moscow, Higher School, 1971.

[48] Abdushukurov, DA; Bondarenko, DV; Chistyakov, DYu. "Calculations of the Efficiency of Registration of Thermal Neutrons by Complex Converters Constructed on the Basis of Gadolinium Foils", *Archives of Los Alamos National Laboratory USA,* 2007, 1-22. http://xxx.lanl.gov/ftp/arxiv/papers/0711/0711.1282.pdf.

[49] X-Ray Data Booklet, *Center for X-ray Optics and Advanced Light Source*, LBNL, USA, http://ie.lbl.gov/atomic/x2.pdf/

[50] Cohen, ER; Taylor, BN. The 1986. *Adjustment of the Fundamental Physical Constants*, CODATA Bulletin 63 (values republished most recently in Physics Today, August 1997, BG7-BG11). 1986.

[51] *X-Ray Mass Attenuation Coefficients,* http://physics.nist.gov/Phys RefData/XrayMassCoef/tab3.html.

[52] Gebauer, B; Alimov, SS; Klimov, A.Yu. et al. Development of hybrid low-pressure MSGC neutron detectors, *Nuclear Instruments and Methods in Physics Research*, 2004, A 529, 358-364.

[53] Paul Csherrer Institute, https://www.psi.ch/niag/

[54] *French Atomic Commission*, http://www-llb.cea.fr/neutrono/nr1.html.

[55] *Hahn-Meitner-Institut*, https://www.helmholtz-berlin.de/zentrum/locations/campus/historie/lise-meitner-campus/index_en.html.

[56] *ISIS*, https://www.isis.stfc.ac.uk/Pages/home.aspx.

[57] *Institute for Energy Technology*, https://www.ife.no/en.

[58] *Frank Laboratory of Neutron Physics Joint Institute for Nuclear Research*, http://flnph.jinr.ru/en/facilities/ibr-2.

[59] *Research Centre REZ*, http://cvrez.cz/en/infrastructure/research-reactor-lvr-15/

[60] Breskin, A; Charpak, G; Majewski, S. On the low pressure operation of multistep avalanche chambers, *Nuclear Instruments and Methods in Physics Research*, 1984, 220, 349.

[61] Abdushukurov, DA; Zanevsky, Yu.V; Movchan, SA. et al. "Multiwire Low Pressure Chamber with a High Gas Amplification Coefficient". *Instruments and Experimental Techniques*, 1983, v. 26, 1287.

[62] Dangendorf, V; Demian, A; Friedrich, H; et al. Thermal neutron imaging detectors combining novel composite foil convertors and gaseous electron multipliers, *Nuclear Instruments and Methods in Physics Research*, 1994, A 350, 503-510.

[63] Abbrescia, M; Iaselli, G; Mongelli, T. et al. Resistive Plate Chambers with Gd-coated electrodes as thermal neutron detectors, *Nuclear Instruments and Methods in Physics Research*, 2004, A 533, 149-153.

[64] Kenji, Takahashi; Seiji, Tazaki; Junji, Miyaharaa. et al. Imaging performance of imaging plate neutron detectors, *Nuclear Instruments and Methods in Physics Research*, 1996, A 377, 119-122.

[65] Gunji, S; Yamashita, Y; Sakurai, H. et al. Development of Neutron Imaging Detector Using Capillary Phenomena and Liquid Scintillator, 2005, *IEEE Nuclear Science Symposium Conference Record*, 2899-2902.

[66] *Neutron imaging – past, present and future,* http://nmi3.eu/news-and-media/neutron-imaging-past-present-and-future.html.

[67] Gilvin, PJ; Matheison, E; Smith, GC. Position resolution in Multiwire Chamber with Graded-Density Cathodes, *IEEE Trans., Nucl. Sci.*, 1981, v. NS-28, No 1, 835.

[68] Abdushukurov, DA; Abduvokhidov, MA; Bondarenko, DV. "Spatial Resolution of the Multistep Avalanche Chambers", *Instruments and Experimental Techniques*, 2007, Vol. 50, No. 3, 333-335.

[69] Abdushukurov, DA; Zanevskii, Yu.V; Peshekhonov, VD. "Effect of gas-mixture composition on characteristics of multistage cascade chambers", *Instruments and experimental techniques*, 1989, Vol. 32, No 1, pt 1, 78-81.

[70] Abdushukurov, DA; Abduvaliev, AA; Abduvokhidov, MA. et al. "Structural Features of the Multistep Avalanche Chambers", *Instruments and Experimental Techniques*, 2007, Vol. 50, No. 1, 37-40.

[71] Breskin, A; Charpak, G; Sauli, F. "*A Multistep Parallax Free Imaging Counter*", CERN-EP/81/106, CERN, 1981.

[72] Altunbas, MC. et al. "Construction, test and commissioning of the triple-GEM tracking detector for COMPASS", *Nucl. Instrum. Meth.* A 490 (2002) 177.

[73] Sauli, F. GEM: "A new concept for electron amplification in gas detectors", *Nucl. Instrum. Meth.* A 386, (1997) 531.

[74] Peisert, A; and Sauli, F. "Drift and Diffusion of Electrons in Gases: A Compilation", *CERN yellow report CERN84-08* (1984).

[75] Bressan et al. Two-dimensional readout of GEM detectors, *CERN-EP*/98-164, (1998).

[76] Dumazert, J; Coulon, R; Lecomte, Q; Bertrand, GNV; Hamel. M. "Gadolinium for neutron detection in current nuclear instrumentation research: A review", *Nuclear Instruments and Methods in Physics Research A,* 2018, DOI: https://doi.org/10.1016/j. (in press).

[77] Bell, ZW; Brown, GM; Ho, CH; Sloop SW. "Organic scintillators for neutron detection", *X-ray and Gamma-Ray Detectors and Applications IV, Proceedings of SPIE,* Vol. 4784, pp. 150-163 (2002).

[78] Aberle, C; Buck, C; Gramlich, B. et al. "Large scale Gd-beta diketonate based organic liquid scintillator production for antineutrino detection", *Journal of Instrumentation,* Vol. 7, pp. 6008- 6023 (2012).

[79] Beil, H; Bergère, R; Veyssière, A. "Détection de photoneutrons avec un scintillateur liquide chargé au gadolinium", *Service de Physique Appliquée,* T. 4, N° 2, pp. 249-250 (1969)

[80] Dazeley, S; Bernstein, A; Bowden, NS; Svoboda, R. "Observation of neutrons with a Gadolinium doped 912 water Cherenkov detector", *Nuclear Instruments and Methods in Physics Research A,* Vol. 607, pp. 616-619 (2009).

[81] Dazeley, S; Sweany, D; Bernstein, A, "SNM detection with an optimized water Cherenkov neutron detector", *Nuclear Instruments and Methods in Physics Research A,* Vol. 693, pp. 148-153 (2012).

Author Contact Information

Djamshed A. Abdushukurov, PhD
Leading Scientific Researcher
Institute of Water Problem, Hydropower and Ecology
Academy of Sciences of Republic of Tajikistan
Dushanbe, Tajikistan
E-mail: abdushukurov.dj@gmail.com

INDEX

I

J

K

L

M

N

P

R

S

T

U

W

X